Me

SUNIL DATTA
Former Professor and Head,
Department of Mathematics and Astronomy
Lucknow University
Lucknow (U.P.)

PHI Learning Private Limited
New Delhi-110001
2010

Rs. 175.00

MECHANICS
Sunil Datta

ISBN-978-81-203-4025-1

The export rights of this book are vested solely with the publisher.

Published by Asoke K. Ghosh, PHI Learning Private Limited, M-97, Connaught Circus, New Delhi-110001 and Printed by Rajkamal Electric Press, Plot No. 2, Phase IV, HSIDC, Kundli-131028, Sonepat, Haryana.

Contents

Preface

In the pages of history of science, Mechanics makes its appearance as first of exact sciences to be developed. Its applications to the development of human resources, and astounding success in explaining analytically and providing quantitative details for the motions of the Earth, the Moon and the other heavenly bodies led to systematic development of science during the last couple of centuries. The source of the remarkable success may be traced to the inlaid beauty of a mathematical discipline embodied in the science of mechanics.

The explanation above affirmatively answers the often raised question–does mechanics merit being included in a Mathematics course? Many of the outstanding contributors to this discipline like–Archimedes, Simon Stevin, Newton, Euler, D'Alembert, Lagrange, Gauss, Stokes, Maxwell, Poincaré, Hamilton, Einstein, Sommerfeld, and Hilbert were mathematicians of par excellence; and this will dispel any remaining doubt.

Classical mechanics may be divided into two branches–Statics and Dynamics. The latter may again be divided into two sub classes: Kinematics, concerning geometry of motion without heeding the force system causing the motion, and Kinetics, concerning the interaction of force and motion. The Greek mathematician Archimedes laid the foundations of Statics in the third century B.C., for giving formulas for simple lever and centre of gravity. About two thousands years later, in the sixteenth century, the Dutch mathematician Simon Stevin gave the parallelogram rule of addition of forces. He also gave the idea of the principle of virtual work, later finalized by John Bernoulli in the eighteenth century. Aristotle tried to explain the behaviour of the moving bodies by erroneously assuming that force was necessary to maintain the motion rather than change its direction and magnitude but he was not succesful. It was left to the genius of Galileo, in the seventeenth century, to disprove the Aristotelians theory and lay the foundations of dynamics. Isaac Newton by inventing calculus, advancing the laws of motion, establishing the gravitational law and explaining the planetary motion, put the science of mechanics on sound and strong mathematical pedestal in the seventeenth century.

This book comprehensively explains forces in three dimensions, orbital motion and particle motion in three dimensions. It digs deep into the subject of space dynamics–launching of satellites and space ships, as it has become is very essential for the students to be exposed in a candid manner with sound mathematical footing. These topics have been included and the vector methods used in a lucid manner, which are lacking in most of the textbooks of this level.

The book consists of ten chapters including Chapter 1 which introduces the reader to the basic concepts, the fundamental laws and the terminology of the science of classical mechanics.

Chapter 2 provides the basic theorems of the three-dimensional force system using the vector method. Chapter 3 enunciates and explains the powerful method of virtual work for the three-dimensional force system. Chapter 4 supplements it by discussing the stability of equilibrium. Chapter 5 studies the equations for strings—common Catenary, strings in contact with curves and so on.

Dynamics part of the book starts with Chapter 6 concentrating on the simplest of motion—the motion in a straight line; it investigates the important cases of S.H.M., motion under inverse square law, resisting medium and motion of a rocket. Chapter 7 provides the basics of kinematics in two dimensions. Chapter 8 exploits the knowledge of the previous chapter to investigate the problems concerning the motion of a particle constrained to move in contact with smooth curves, particularly the circular motion and the cycloidal motion. Chapter 9 investigates the orbital motion under a central force system—Planetary motion, Rutherford scattering, and motion of artificial satellites. The last Chapter 10 studies through vector methods the motion of a particle in three-dimensions elucidating the kinematics in spherical polar, cylindrical polar and in natural co-ordinates. The knowledge leads to study the motion in a rotating frame of reference particularly the important case of rotating earth.

The book contains adequate number of suitable examples elucidating and illustrating the basics to make the student competent enough to understand and tackle the exercises accompanying each of the chapters.

The book has emerged out of the lecture notes that I had the pleasure of imparting to my undergraduate students for a number of years at Lucknow University. It has been written primarily for mathematics undergraduates of Indian Universities but others including engineering students stand to gain from it. In a classical subject like mechanics nothing is new except for the approach which the reader will find quite deviating from the contemporary textbooks. It is meant for students who wish to encounter and overcome difficulties, thereby enhancing his understanding and appreciation of the discipline.

The book has been designed to cater to the needs of a student who wishes to encounter and overcome difficulties. The introducing of the question mark (?) symbol at various places in the text is to induce the student to think, work and assimilate.

Needless to say that this book has drawn from various classical standard books on the subject and I wish to acknowledge my thanks to all such sources. I also extend my thanks to Prof. Kamla D. Singh, former Head of the department of Mathematics and Astronomy, Lucknow University for her encouragement, and Prof. Arun Verma of IIT Roorkee for the help rendered to me in the preparation of this book.

Sunil Datta

Mechanics–Basic Concepts

1.1 INTRODUCTION

Nature manifests itself in the form of matter. The earth, the stars and the galaxies are all material bodies. Be it a man or a machine, both living and non-living world are composed of some or other type of matter. An intrinsic property of matter is its mass as appearing in Newton's law of gravitation, which is in the form of weight in humans. Electric charge and temperature may be cited as other examples of basic attributes of physical bodies. In the restless cosmos, things change in course of time. The simplest of the change is the change in the position with respect to some co-ordinate system. This shifting introduces the concept of motion. Mechanics is the discipline dealing with matter in conjunction with motion and time. It is the branch of physics dealing with the action of force on material bodies, a branch of applied mathematics treating motion; and also the science of machinery. With the advancements of knowledge, mechanics came to encompass the kingdoms of thermodynamics and electrodynamics. In this book we confine our attention to the classical mechanics or Newtonian mechanics as it is commonly known. In the study of mechanics, it may happen that the system being studied is at rest — is static, then study is termed statics. In statics we take the body to be perfectly rigid and at rest implying that the action of all the forces acting on it is null and that there is no tendency for the forces to move and turn the body. On the other hand, when the object is in motion changes take place continuously, we have the case of Dynamics which may itself be subdivided into *kinematics* and *kinetics*. In kinematics, we study motion without considering forces involved, and in kinetics we study the relation between forces and motion. Matter comes into play in numerous forms and in wide variety of structures. But broadly speaking it is found to occur in following four states — solid, liquid, gas and plasma. Matter is composed of molecules. The same substance may occur in different state, e.g., ice, water and steam. The state depends on the looseness of the bonds between the molecules. We can loosen the bonds by supplying heat, thereby increasing the temperature. The fourth state of matter plasma, though most widely, occurring (99 per cent of our Universe is made up of plasma) does not form a part of classical mechanics as in its very conception plasma is made up of electrically charged particles.

Mass is an intrinsic quantity determined by the content of the matter associated with a body. Matter in its first three states, viz., solid, liquid and gas is composed of electrically neutral molecules and the ideal aim of mechanics will be to study the motion of individual molecules. However, it is a bit difficult to study all the properties because of their numerical abundance. Therefore, instead, we contemplate the matter to be continuously distributed, i.e., it can be treated as a continuum; thereby introducing the concept of mass density

$$\rho = \lim_{\Delta v \to 0} \frac{\Delta m}{\Delta v}, \tag{1.1}$$

where Δm is the mass and Δv is the volume at the field position.

Here it should be kept in mind that although $\Delta v \to 0$, in practical situation Δv, although small, will be large enough to still hold a multitude of molecules because of their extreme smallness in size. Thus, a tiny bit of material body possessing mass, but point linear dimensions will be termed a particle. But when studying motion on atomic scale the individual particle character cannot be ignored and inter atomic forces over looked. We have then to take recourse to quantum mechanics. Thus, the concepts of classical mechanics are not adequate for studying the microscopic structure of matter. However, when very large masses, very high speeds or very long distances are involved, the proper framework to deal with the problem is that of relativistic mechanics.

Mechanics is broadly divided into two parts:

(i) Statics: dealing with equilibrium of bodies at rest.

(ii) Dynamics: dealing with bodies in motion. This may again be divided into two sub classes—(a) Kinematics which deals with geometry of motion without heeding the force system causing the motion, and (b) Kinetics which deals with interaction of force and motion.

1.2 NEWTON'S LAWS OF MOTION

The foundations of Newtonian Mechanics are the three laws of motion, which were propounded by Newton in his celebrated publication *Principia*, which appeared in 1687. Before enunciating these laws we define below some basic quantities; the student may already be familiar with these. If $\Delta\mathbf{r}$ is the displacement in time Δt in the position of the particle which at time t occupies the position $\mathbf{r}$, then its velocity is given by

$$\mathbf{v} = \lim_{\Delta t \to 0} \frac{\Delta \mathbf{r}}{\Delta t} = \frac{d\mathbf{r}}{dt}.$$

It may be noted that for the movement of a point only a single vector $\mathbf{v}$ suffices, but when we consider the motion of a straight line, as in the case of a rigid body, we have to prescribe its translation (moving parallel to itself) as well as its rotation.

The acceleration $\mathbf{f}$ of a particle is the rate of change of its velocity

$$\mathbf{f} = \frac{d\mathbf{v}}{dt} = \frac{d^2\mathbf{r}}{dt^2}.$$

With m as the mass of the particle, we define its momentum as

$$\mathbf{p} = m\mathbf{v}. \tag{1.2}$$

The momentum $\mathbf{P}$ of a body is the sum of the momentum of individual particles composing it. Hence, we get

$$\mathbf{P} = \Sigma m\mathbf{v} = \frac{d}{dt}\Sigma m\mathbf{r}. \tag{1.3}$$

Then, the angular momentum of a particle at $\mathbf{r}$ about the origin O is mathematically defined as

$$\mathbf{h} = \mathbf{r} \times \mathbf{p} = m\,\mathbf{r} \times \mathbf{v}, \tag{1.4}$$

i.e., it is the moment of momentum. However, the angular momentum of a body is the vector sum of the angular momentum of individual particles composing it

$$\mathbf{H} = \Sigma m\mathbf{r} \times \mathbf{v}. \tag{1.5}$$

Suppose a point P is rotating about an axis directed along unit vector $\hat{\mathbf{a}}$; then it remains in a circle lying on a plane perpendicular to $\hat{\mathbf{a}}$. The centre O of the circle obviously lies on the axis. Let θ be the angle which the line OP makes with a fixed line in the plane of the circle. Let $\omega = \frac{d\theta}{dt}$ be the time rate of change of the angle θ. Then, the vector $\boldsymbol{\omega} = \left(\frac{d\theta}{dt}\right)\hat{\mathbf{a}}$ is called the angular velocity of point P about the axis. In the case of uniplanar motion, ω itself may be called the angular velocity of the point P or of line OP. More generally, it may be seen that the angular velocity of a line P_1P_2 moving in a plane is

$$\frac{(v_{2n} - v_{1n})}{|\mathbf{P}_1\mathbf{P}_2|},$$

where v_{1n} and v_{2n} are the components of its end point velocities along the direction perpendicular to P_1P_2. If ψ is the slope of a tangent, then $\frac{d\psi}{dt}$ is called the *angular velocity* of the tangent.

Now, we present the *three* Newton's laws of motion.

Newton's first law of motion

Newton's first law states that *a body continues its state of rest or of uniform motion in a straight line unless or until acted upon by a force.* It introduces the force as an agency which changes or tends to change the state of motion of a body. The net force **F** on a body is the sum of forces acting over discrete points of the body or may be expressed through the integral of a continuous force density vector. Engineers prefer to recognize the force through the deformation that it imparts to a material body. Thus, force may be defined by the amount of elongation undergone by a spring caused by it.

Newton's second law of motion

Newton's second law of motion states that *the rate of change of momentum of a body is directly proportional to the external force applied to it.* The second law tells us about the change brought up by the force in the state of a body. Analytically, it may be represented as

$$\frac{d\mathbf{P}}{dt} = \mathbf{F}, \tag{1.6}$$

where unit of force is conveniently taken to render the constant of proportionality have unit value. If the body is a particle of constant mass m the above Eq. (1.6) reduces to

$$\frac{d\mathbf{P}}{dt} = m\frac{d\mathbf{v}}{dt} = \mathbf{F} \tag{1.7}$$

or

$$m\frac{d^2\mathbf{r}}{dt^2} = \mathbf{F}. \tag{1.8}$$

Equation (1.6) is the mathematical embodiment of the second law of motion for the linear momentum. But it is easy to conceive that in the case of a body, besides linear motion there may also be rotation, and then we have to write down the corresponding equations for angular momentum **H** which may be derived from Eq. (1.6) as

$$\frac{d\mathbf{H}}{dt} = \mathbf{M}, \tag{1.9}$$

where **M** is the vector moment of the external forces about O [see Chapter 2]. But since this course is confined to the dynamics of a particle, we shall not probe further into the realm of rigid body dynamics.

Newton's third law of motion

Newton's third law of motion states that to *every action there is an equal and opposite reaction.* The third law introduces the force of reaction and is vital in generating the constraint forces in problems in mechanics.

We may introduce through Newton's second law inertial mass and is identical to the gravitational mass as introduced through Newton's gravitational law of attraction between bodies.*

1.3 FRAME OF REFERENCE

Newton considered time and space as distinct quantities. According to classical mechanics, time, position and motion are relative to a frame of reference, which forms an event world. Thus, a co-ordinate system (which may be taken as a Cartesian-rectangular) together with a clock constitute a frame of reference and an event is said to take place at certain place in space at certain time, that is how we describe the happenings in this Universe—be it the Halley's comet making its periodic appearance, a rocket speeding to moon, a locomotive steaming into a platform or a mosquito disturbing your sleep. There is no harm, if we set all our watches at the same time and ignore its translation. Thus, the physical space is taken as Euclidean and time uniquely and identically defined for all its points. Amongst the multitude of spatial co-ordinates, the question naturally arises, whether there is any special absolute co-ordinate system or not? The answer, based on experimental observations is that there is a special class named *inertial frame* or the *Newtonian frame*, and in that alone a body continues its state of rest or of uniform motion until a force is applied. There is no one single absolute inertial frame, but if we know one then all the others moving with uniform velocity relative to it are inertial without distinction as far as dynamics is concerned. Thus, we can have a frame fixed in a train moving with constant speed over a straight track so that a coin rests on its edge without falling down. For example, if all the windows are closed in a train we can sit comfortably on the moving train without realizing that we are not sitting at our home. However the situation is not the same when the train approaches a station or deaccelerates to come to rest or when it negotiates a bend. At that time we experience a force and have to hold on to something to avoid tumbling. In the solar system, a frame of reference in which the centre of mass of the system is fixed with respect to certain distant stars have no discernible motion can be taken as the inertial frame. Because of the dominant mass of the Sun, the centre of the mass can be deemed to be at the centre of the Sun, which is the Sun-star frame of reference. For most of the problems near the surface of the earth, like the study of motion of an artificial satellite, this frame of reference suffices. An earth-star frame with the centre of the earth as fixed point with respect to distant stars may also be adequate for dealing with practical problems on the earth's surface. In most of the engineering problems and laboratory problems near the surface of the earth a frame of reference

---

*[For an excellent discussion of the experiments on the relation of two kinds of masses see: C.R.H. Dike, Gravitation and the Universe, (1970), American Philosophical Society, Philadelphia].

attached to a point on the surface will be sufficient because the inertial forces arising out of the movement of the earth around the sun or its rotation about its axis are negligible.

But for investigating intergalactic problems some other frame is needed. It is important to bear in mind that it is physical experience that leads us to determine whether a particular reference frame can be taken as inertial or not. Further, it should be noted that although the use of inertial frame is convenient generally, but sometimes non-inertial frames like rotating frames may also be found useful. Also, observe that in all imaginable inertial frames acceleration will be same. In inertial frames, the clocks are normally taken to be synchronized so that the same time t prevails in all frames, and then transformation of frames is same as transformation of co-ordinate systems.

Thus, armed with a reference frame (or co-ordinate system), we have the position of a particle of mass m given by its position vector $\mathbf{r}$ which according to Newton's second law changes, as described by the second order vector differential Eq. (1.8). The agency of change is force $\mathbf{F}$, and a few words about it will not be out of place.

True forces in nature are of three kinds: gravitational forces, electromagnetic forces, and nuclear forces. Gravitational forces are relatively very weak, but electromagnetic forces are of intermediate strength, and are perceived directly by people in the macroscopic world. Such forces as the force of friction, the elastic force in a spring, the tension in a rope, the force exerted by muscles, are all nothing but macroscopic manifestations of the electromagnetic forces. Nuclear forces are of two types, i.e. strong and weak, but they do not play any role in classical mechanics.

1.4 WORK, ENERGY AND IMPULSE

The work done by a force $\mathbf{F}$ acting on a particle as the particle gets displaced from a point P_1 to point P_2 is defined as

$$W = \int_{P_1}^{P_2} \mathbf{F} \cdot d\mathbf{r}, \tag{1.10}$$

where $d\mathbf{r}$ is the line element of the path C along which the line integral is taken. In general, W depends on the path C. But if the line integral is independent of the path the force field is said to be conservative. For a conservative force field, the line integral around a closed path is zero, i.e.

$$\oint \mathbf{F} \cdot d\mathbf{r} = 0, \tag{1.11}$$

and conversely.

Earth's force of gravity $\mathbf{g}$ and central force $\mathbf{F} = F(r)\,\hat{\mathbf{r}}$ are examples of conservative force field. Students may refer to their course on vector calculus to conclude that for a force field $\mathbf{F}$ satisfying Eq. (1.11) equivalently

$$\text{curl } \mathbf{F} = 0 \tag{1.12}$$

or that there exists a potential function V such that

$$\mathbf{F} = -\text{ grad } V. \tag{1.13}$$

The above result provides

$$V = -\int_{P_0}^{P} \mathbf{F} \cdot \mathbf{dr} \tag{1.14}$$

which may be interpreted as the work done against the force in moving the particle from some standard position P_0 (which may be taken at infinity) to the field position P. This may be taken as the definition of the potential energy (P.E.) V. Thus, a particle acquires potential energy by virtue of its position. If V_1 and V_2 are potential energies at positions P_1 and P_2, respectively, then it follows from Eq. (1.14) that

$$\int_{P_1}^{P_2} \mathbf{F} \cdot \mathbf{dr} = V_1 - V_2. \tag{1.15}$$

Remark: It may be noted here that *energy is a scalar quantity* and may be interpreted as the capacity of work done by forces. If the forces do work on a system, the energy of the system increases, but if the forces of the system do work then energy of the system is consumed. Students will find it no difficult to work out that in force of gravity g, the potential energy increases by mgh when the particle goes up a height h, and the potential energy stored in an elastic string is the product of mean tension and extension [see Example 1.6].

Next, using Eq. (1.7), we can also write

$$\int_{P_1}^{P_2} \mathbf{F} \cdot \mathbf{dr} = m\int_{P_1}^{P_2} \frac{\mathbf{dv}}{dt} \cdot \mathbf{dr} \tag{1.16}$$

$$= m\int_{P_1}^{P_2} \mathbf{v} \cdot \mathbf{dv} \qquad \left(\text{since } \frac{dr}{dt} = \mathbf{v}\right)$$

$$= \frac{1}{2} m (v_2^2 - v_1^2).$$

The quantity $K = \frac{1}{2} mv^2$ is termed as the kinetic energy (K.E.) of a particle of mass m moving with speed v. Equation (1.16) embodies the *work energy theorem* viz., the work done on a particle is equal to increase in its kinetic energy. The work energy theorem is simply the first integral of the equation of motion, and does not represent a law independent of Newton's second law. It is helpful in obtaining the speed when the work done is easily computable. Moreover, its importance lies in the fact that it leads to far reaching developments in physics.

Now, if the force field is conservative, from Eqs. (1.15) and (1.16), we have

$$\frac{1}{2}mv_2^2 + V_2 = \frac{1}{2}mv_1^2 + V_1 = E \tag{1.17}$$

where the mechanical energy E is a constant. Equation (1.17) gives the mathematical expression for the principle of *conservation of energy*, i.e. for a particle moving in a conservative force field, the sum of kinetic and potential energies is constant. This implies that an increase in K.E. is equal to a decrease in P.E. It may also be noted that the principle of conservation of energy also holds good for a system of particle in a conservative force field.

Equation (1.17) may be extended to include other kinds (heat, electromagnetic, etc.) of energies to lead to the general principle of conservation of energy viz., the total energy of a system is constant; it cannot be created or destroyed but may be transferred from one kind to the other. The energy principle is one of the unifying ideas in physics. We must keep in mind that kinetic and potential energies sum up to a constant only when conservative forces act, but in the general, the total energy is always a constant. One should understand that when frictional force **f** is present in an otherwise conservative system, the rate at which mechanical energy is dissipated is given by the equation

$$\frac{d}{dt}(K + V) = -\mathbf{f} \cdot \mathbf{v} \tag{1.18}$$

where **v** is the velocity at that instant.

We have seen that the space integral of the force leads to kinetic energy. Similarly, the time integral of the force defines impulse (**I**) as follows:

$$\mathbf{I} = \int_{t_1}^{t_2} \mathbf{F} dt = m \int_{t_1}^{t_2} d\mathbf{v} = m(\mathbf{v}_2 - \mathbf{v}_1) = \mathbf{p}_2 - \mathbf{p}_1. \tag{1.19}$$

In case the duration of the impulse (the time interval $t_2 - t_1$) is infinitesimal, but the force is large enough so that

$$\mathbf{I} = \lim_{t_2 \to t_1} \mathbf{F}(t_2 - t_1)$$

is finite, we have the case of impulsive force. Thus, impulsive force is determined by the instantaneous change in momentum.

Also, we find that

$$\text{Increase in K.E.} = \frac{1}{2}m(\mathbf{v}_2^2 - \mathbf{v}_1^2)$$

$$= \frac{1}{2}m(\mathbf{v}_2 - \mathbf{v}_1)(\mathbf{v}_2 + \mathbf{v}_1) = \mathbf{I} \cdot \mathbf{v}_m. \tag{1.20}$$

where $\mathbf{v}_m = \frac{1}{2}(\mathbf{v}_1 + \mathbf{v}_2)$ is the mean velocity.

Student will not find it difficult to understand that if a ball of mass m strikes a wall normally with speed u, and rebounds without change in the speed, then the average force exerted by the ball on the wall is $\frac{2mu}{\Delta t}$, where Δt is the time of collision.

When there is no impulse $\mathbf{I} = 0$, and we have from Eq. (1.19)

$$\mathbf{p}_2 = \mathbf{p}_1, \tag{1.21}$$

which embodies the principle of conservation of linear momentum. It also follows directly from Eq. (1.6) by taking $\mathbf{F} = 0$ and thus getting $\mathbf{P}$ = Constant.

1.5 RELATIVE MOTION

The position vector of a point in space is linked with the co-ordinate system under consideration. Let a point P in space have the position vector $\mathbf{r}$ with respect to a co-ordinate system S with origin O and the position vector $\mathbf{r}'$ with respect to a co-ordinate system S′ with origin O′, i.e.

$$\mathbf{OP} = \mathbf{r}, \text{ and } \mathbf{O'P} = \mathbf{r}'.$$

With $\mathbf{OO'} = \mathbf{s}$, we have

$$\mathbf{r} = \mathbf{s} + \mathbf{r}'.$$

On differentiation with respect to time, we obtain

$$\mathbf{v} = \mathbf{u} + \mathbf{v}', \tag{1.22}$$

where $\mathbf{v}$ is the velocity of P with respect to the O (co-ordinate system S, which may be considered to be fixed), $\mathbf{v}'$ with respect to the O' (moving co-ordinate system S′) and $\mathbf{u}$ the velocity with which O' moves with respect to O (the translation velocity of S′). We may express Eq. (1.22) as

$$\mathbf{v}' = \mathbf{v} - \mathbf{u}. \tag{1.23}$$

Hence, the velocity in a moving frame is obtained from the fixed frame velocity by subtracting the velocity of the moving frame.

Equation (1.23), on differentiation, provides for the acceleration of as point P in the moving frame

$$\mathbf{f}' = \mathbf{f} - \mathbf{a}, \tag{1.24}$$

$\mathbf{a}$ being the acceleration of the translating frame.

Consider transformation of co-ordinates from one inertial system S to another S′. The clocks being synchronized so that $t' = t$. We can conveniently take the uniform velocity along x direction, providing $\mathbf{s} = ut\mathbf{i}$, and hence we have the *Galilian transformation* as

$$\mathbf{r}' = \mathbf{r} - u\,t\mathbf{i},\ t' = t \tag{1.25}$$

or in Cartesian co-ordinates

$$x = x' - ut,\ y' = y,\ z' = z,\ t' = t. \tag{1.26}$$

In the special theory of relativity, taking speed of light c to be invariant of the translating frame, the appropriate transformation is the Lorentz transformation

$$Tx = x' - ut, y' = y, z' = z, T t' = t - \frac{ux}{c^2}, \tag{1.27}$$

where $T^2 = 1 - \frac{u^2}{c^2}$. It may be noticed that two clocks in the fixed and moving frames are bound to register different times. Also, Newtonian mechanics holds good only when the speeds involved are much smaller than the speed of light.

1.6 NON-INERTIAL FRAMES

Newton's laws of motion are applicable only in inertial frames. In non-inertial frames, the laws are to be modified to take into account the fictitious forces arising out of the acceleration of the reference frame in which the measurements are made. For the linear case, we have the examples of the motion relative to a lift or an accelerating car on a straight road. [See Example 1.10]

With the reference frame attached to a turntable, e.g., a record player, a spin dryer, a CD player or rotating earth, forces such as centrifugal force, coriolis force appear. These are fictitious forces (or pseudo-forces). They may also be termed as inertial forces as they arise when we choose a rotating frame as a reference frame. We shall study them in Chapter 10.

SOLVED EXAMPLES

EXAMPLE 1.1 If the force of attraction between a pair of particles of masses m_1 and m_2 is given by $\frac{km_1m_2}{x^2}$, where k is a constant and x the distance between the particles, find

(i) the potential energy

(ii) the work required to increase the separation from $x = x_1$ to $x = x_2$.

Solution

(i) We have

$$V = -\int_{\infty}^{x} \mathbf{F} \cdot dr = km_1m_2 \int_{\infty}^{x} \frac{dx}{x^2} = -\frac{km_1m_2}{x}.$$

(ii) Work required to increase the separation

$$W = km_1m_2 \int_{x_1}^{x_2} \frac{dx}{x^2} = V_2 - V_1 = \frac{km_1m_2}{x_1x_2}(x_2 - x_1).$$

EXAMPLE 1.2 A flexible but inextensible chain of length l and mass ml is held on a smooth table with the length $l - a$ on the table and the length a overhanging. Find the velocity with which the chain will leave the table.

Solution Since the initial speed is zero and whole of the chain is moving, the increase in K.E. is given by

$$E_1 = \frac{1}{2} mlv^2 \tag{i}$$

where v is the speed with which the chain leaves the table.

Next, the weight of the overhanging part of the chain can be taken at the centre of gravity of it. Since the length of the overhanging part increases from the value a to final value l, the decrease in P.E. is given by

$$E_2 = mgl\frac{l}{2} - mga \cdot \frac{a}{2} = \frac{1}{2} mg(l^2 - a^2)$$

Now, from the principle of conservation of energy, we have

Increase in K.E. = Decrease in P.E.

or
$$\frac{1}{2} mv^2 = \frac{1}{2} mg(l^2 - a^2)$$

providing
$$v = \sqrt{\frac{g}{l}(l^2 - a^2)}.$$

EXAMPLE 1.3 A body of mass m starts from rest down a plane of length l inclined at an angle α with the horizontal.

(i) Find the body's speed at the bottom taking the coefficient of friction as μ.

(ii) How far will it slide horizontally on a similar surface after reaching the bottom of the incline?

Solution

(i) The force of friction on the inclined plane given by

$$F = \mu R = \mu mg \cos\alpha$$

is seen to be a constant, and is opposite to the motion. The body moves down a distance l along the plane, i.e. a vertical distance $l \sin \alpha$. Therefore, if v is body's speed at the bottom, we have from work energy theorem as

$$\frac{1}{2} mv^2 - mgl \sin\alpha = -\mu mg \cos\alpha \cdot l,$$

providing
$$v = \sqrt{\{2gl(\sin\alpha - \mu\cos\alpha)\}}.$$

(ii) If d is the distance moved on the horizontal plane before the body comes to rest, we again have from work energy theorem

$$\mu mg\, d = \frac{1}{2} mv^2 = mgl(\sin\alpha - \mu\cos\alpha),$$

giving $$d = \frac{l(\sin\alpha - \mu\cos\alpha)}{\mu}.$$

EXAMPLE 1.4 If an elastic string, whose natural length is that of a uniform rod, is attached to the rod at both ends and suspended by the middle point, show that the rod will sink until the strings are inclined to the horizontal at an angle θ given by the equation

$$\cot^3\frac{\theta}{2} - \cot\frac{\theta}{2} = 2n,$$

given that the modulus of the string is n times the weight of the rod.

Solution Here, as the rod goes down, the P.E. gets stored as the elastic energy in the string. Let us suppose that the length of the rod and the initial length of the string is $2\,l$. When in lower position the length of the string will be $2l \sec\theta$, the elastic energy stored in it will be given by [using Example 1.6]

$$E_1 = \frac{1}{2l} nW(2l\sec\theta - 2l)(2l\sec\theta - 2l). \qquad \text{(i)}$$

Also, since the centre of gravity of the rod is lowered by a distance $l\tan\theta$, the decrease in P.E. is

$$E_2 = Wl\tan\theta. \qquad \text{(ii)}$$

The equation $E_2 = E_1$ will lead to the desired result. (The students are advised to draw a figure.)

EXAMPLE 1.5 An electron of mass m collides head on with an atom of mass M which was initially at rest. As a result of this collision a characteristic amount of energy E is stored internally in the atom. What is the minimum initial velocity u_0 that the electron must have?

Solution Suppose u and U are the final velocities of the electron and the atom, respectively. From momentum conservation, we have

$$mu + MU = mu_0 \qquad \text{(i)}$$

From energy principle, we get

$$\frac{1}{2}mu^2 + \frac{1}{2}MU^2 + E = \frac{1}{2}mu_0^2. \qquad \text{(ii)}$$

Eliminating U from Eqs. (i) and (ii), we have the quadratic equation for u as

$$\left(1 + \frac{m}{M}\right)u^2 - \frac{2m}{M}u_0 u + 2E - \left(1 - \frac{m}{M}\right)u_0^2 = 0$$

providing

$$u = \frac{m}{M}u_0 \pm \sqrt{\left[u_0^2 - \frac{2(m+M)E}{mM}\right]}.$$

Now, for a real value of u, it is necessary that

$$u_0^2 > \frac{2(m+M)E}{mM}$$

and this provides the minimum value of u_0 as $\left[\frac{2(m+M)E}{mM}\right]^{\frac{1}{2}}$.

EXAMPLE 1.6 Show that the energy stored in an elastic string is given by the product of mean tension and extension.

Solution Let an elastic string, obeying Hooke's law, of natural length l and modulus of elasticity λ be extended from a length l_1 ($>l$) to a length l_2 with tensions having respective values T_1 and T_2.

Now, for extended length x, tension $T = \frac{\lambda x}{l}$; hence, the work done on the system by the tension T, i.e., increase in the energy $\delta E = T\delta x$. Thus, the total *elastic energy stored in the string* is

$$E = \int_{x1}^{x2} Tdx$$

$$= \int_{x1}^{x2} \frac{x}{l}\, dx$$

$$= \lambda \frac{(x_2 + x_1)(x_2 - x_1)}{2l}$$

$$= \frac{1}{2}(T_2 + T_1)(x_2 - x_1).$$

EXAMPLE 1.7 Show that the elastic force in a spring is conservative.

Solution Let l be the natural length of an elastic string and k its stiffness. Let its one end be fixed and the other end P move on a closed curve C so that $\text{OP} = \mathbf{r} = r\,\hat{\mathbf{r}}$, the stretched length being r. The force of tension in the spring is

$$\mathbf{F} = -\,k(r-l)\hat{\mathbf{r}}.$$

Therefore,

$$\oint_C \mathbf{F}\cdot d\mathbf{r} = -\frac{k}{l}\oint_C (r-l)\hat{\mathbf{r}}\cdot d\mathbf{r}$$

$$= -\frac{k}{l}\oint_C (r-l)\,dr = 0 \quad (\text{since } \mathbf{r}\cdot d\mathbf{r} = rdr)$$

showing that the force field is conservative.

EXAMPLE 1.8 Two ships A and B start sailing at the same time from positions **a** and **b** relative to an origin O with uniform velocities **u** and **v**. Determine the shortest distance between the ships and the time at which this occurs.

Solution The positions of the two ships at any time t are given by

$$\mathbf{r} = \mathbf{a} + \mathbf{u}t \text{ and } \mathbf{s} = \mathbf{b} + \mathbf{v}t.$$

The distance D between the two ships at any time t is given by

$$D^2 = (\mathbf{r}-\mathbf{s})\cdot(\mathbf{r}-\mathbf{s}) = (\mathbf{a}-\mathbf{b})^2 + 2(\mathbf{a}-\mathbf{b})\cdot(\mathbf{u}-\mathbf{v})t + (\mathbf{u}-\mathbf{v})^2 t^2,$$

providing
$$D\frac{dD}{dt} = (\mathbf{u}-\mathbf{v})\cdot[(\mathbf{a}-\mathbf{b}) + (\mathbf{u}-\mathbf{v})t].$$

Therefore, for a maxima or minima, the vanishing of the derivative provides the time $t = T$ as

$$T = \frac{(\mathbf{v}-\mathbf{u})\cdot(\mathbf{a}-\mathbf{b})}{(\mathbf{u}-\mathbf{v})^2}, \ (\mathbf{u} \neq \mathbf{v}).$$

Next, we find that

$$\left(D\frac{d^2D}{dt^2}\right)_{t=T} = (\mathbf{u}-\mathbf{v})^2. \tag{i}$$

As $D > 0$, this implies that

$$\left(\frac{d^2D}{dt^2}\right)_{t=T} > 0. \tag{ii}$$

Hence, we have a minima.

Using value of T, we have the shortest distance between the ships given by

$$D_{\min} = \left[(\mathbf{a}-\mathbf{b})^2 - \frac{\{(\mathbf{u}-\mathbf{v})\cdot(\mathbf{a}-\mathbf{b})\}^2}{(\mathbf{u}-\mathbf{v})^2}\right].$$

EXAMPLE 1.9 An airplane whose speed is u in still air flies horizontally along a square path ABCD of a side of length c, and AB and DC parallel to south north direction. If the wind is blowing with speed v from a direction β east of north, show that the total time taken to complete the circuit ABCD is

$$\frac{2c}{u^2 - v^2}[(u^2 - v^2\sin^2\beta)^{\frac{1}{2}} + (u^2 - v^2\cos^2\beta)^{\frac{1}{2}}].$$

Solution With D as origin and DA and DC being along east and north directions, let the points A, B and C be $c\mathbf{i}$, $c(\mathbf{i} + \mathbf{j})$ and $c\mathbf{j}$, respectively.

The wind velocity may be expressed as [See Figure 1.1]

$$\mathbf{v} = -\, v(\sin\beta\,\mathbf{i} + \cos\beta\,\mathbf{j}).$$

Let us suppose that the airplanes velocity relative to wind is

$$\mathbf{u} = u(\cos\theta\,\mathbf{i} + \sin\theta\,\mathbf{j}),$$

its speed being u, for moving along the path AB in the presence of wind. If its speed is w. we have

$$\mathbf{w} = \mathbf{u} + \mathbf{v} \tag{i}$$

or
$$w\mathbf{j} = u(\cos\theta\,\mathbf{i} + \sin\theta\,\mathbf{j}) - v(\sin\beta\,\mathbf{i} + \cos\beta\,\mathbf{j}), \tag{ii}$$

providing
$$u\cos\theta - v\sin\beta = 0$$

and
$$u\sin\theta - v\cos\beta = w$$

Eliminating θ, we get the equation for w as

$$w^2 + 2v\,w\cos\beta + v^2 - u^2 = 0. \tag{iii}$$

Providing
$$w = \left[\pm(u^2 - v^2\sin^2\beta)^{\frac{1}{2}} - v\cos\beta\right]^{\frac{1}{2}}.$$

Obviously (?) + sign is the correct one (and also $u > v$?) then the time $t_1 = \dfrac{c}{w}$ to cover AB is

$$t_1 = \frac{c}{u^2 - v^2}\left[(u^2 - v^2\sin^2\beta)^{\frac{1}{2}} + v\cos\beta\right]. \tag{iv}$$

Alternately, Eq. (iii) can be easily derived from Eq. (i) by expressing it as $\mathbf{w} - \mathbf{v} = \mathbf{u}$, and observing that β is the angle between the directions of $\mathbf{w}$ and $-\mathbf{v}$.

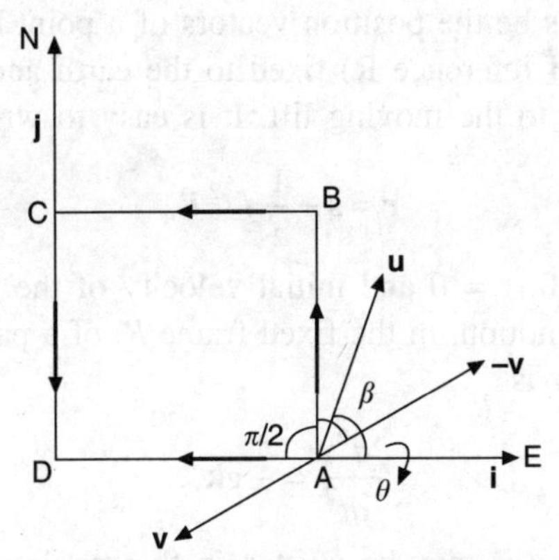

Figure 1.1 Airplane flying along the square path ABCD.

Noting that the angle between $\mathbf{w}$ and $-\mathbf{v}$ for motion along BC is $\dfrac{\pi}{2} + \beta$, the time t_2 for covering BC may be obtained from Eq. (iv) by replacing β by $\dfrac{\pi}{2} + \beta$, thus, we have

$$t_2 = \frac{c}{u^2 - v^2}\left[(u^2 - v^2\cos^2\beta)^{\frac{1}{2}} - v\cos\beta\right].$$

Again, for motions along CD and DA, the corresponding angles are $\pi - \beta$ and $\frac{\pi}{2} - \beta$ and these provide the corresponding times t_3 and t_4 for traversing CD and DA as

$$t_3 = \frac{c}{u^2 - v^2}\left[(u^2 - v^2\sin^2\beta)^{\frac{1}{2}} - v\cos\beta\right]$$

and
$$t_4 = \frac{c}{u^2 - v^2}\left[(u^2 - v^2\cos^2\beta)^{\frac{1}{2}} + v\sin\beta\right].$$

Hence, the total time

$$T = t_1 + t_2 + t_3 + t_4$$

$$= \frac{2c}{u^2 - v^2}\left[(u^2 - v^2\sin^2\beta)^{\frac{1}{2}} + (u^2 - v^2\cos^2\beta)^{\frac{1}{2}}\right].$$

EXAMPLE 1.10 Consider the case of a lift moving upwards in the unit direction **k** with uniform acceleration *f*. Show that a man in the lift experiences an additional force.

Solution Let **r** and **s** be the position vectors of a point P in space relative to an origin O (frame of reference R) fixed to the earth and an origin Q (frame of reference S) fixed to the moving lift. It is easy to write the relation

$$\mathbf{r} = \mathbf{s} + \frac{1}{2} f t^2\,\mathbf{k}, \tag{i}$$

assuming that at $t = 0$, $\mathbf{r} = 0$ and initial velocity of the lift is 0.

The equation of motion, in the fixed frame R, of a particle of mass m and moving under gravity is

$$\frac{d^2\mathbf{r}}{dt^2} = -g\mathbf{k}, \tag{ii}$$

and this on using Eq. (ii) may be written in the moving frame as

$$\frac{d^2\mathbf{s}}{dt^2} = -g\mathbf{k} - f\,\mathbf{k}. \tag{iii}$$

This is the reason that a man in an upward accelerating lift feels an additional force.(?)

EXAMPLE 1.11 Find whether the following velocity dependent forces are conservative:

(i) Friction force, $\mathbf{F} = -c\mathbf{v}$

(ii) Lorentz force on a charge q moving in magnetic field $\mathbf{B}$, $\mathbf{F} = q\,\mathbf{v} \times \mathbf{B}$

Solution We shall evaluate the line integral $\mathrm{I} = \oint \mathbf{F} \cdot \mathbf{dr}$ round a closed circuit to check the nature of the force.

(i) $\mathrm{I} = -c\oint \mathbf{v} \cdot \mathbf{dr} = -c\oint \mathbf{v} \cdot \mathbf{v}dt = -\frac{1}{2}c\oint \mathbf{v}^2 \mathrm{d}t \neq 0.$

Hence, frictional force is not conservative.

(ii) $\mathrm{I} = q\oint \mathbf{v} \times \boldsymbol{B} \cdot \mathbf{dr} = q\oint (\mathbf{v} \times \boldsymbol{B}) \cdot \mathbf{v}\,\mathrm{d}t = 0.$

Hence, Lorentz force is conservative.

PROBLEMS

1. Two particles P and Q are connected by a light string of length l, and rest on a smooth table. If the string is just taut when P is struck with an impulse I at right angles to PQ, describe the subsequent motion if P has mass m and Q has a mass of $2m$.

[**Ans:** Mass centre moves perpendicular to PQ with speed I/3 m); AB rotates about mass centre with angular speed (I/lm)]

2. A rocket of mass m moves vertically upward with constant acceleration f starting from its state of rest on the surface of the earth. The force of attraction $F(x)$ is inversely proportional to the square of the distance x from the body to the centre of the earth; the air resistance is neglected. Determine the thurst force T making the rocket move upward and the work W of this force performed as the rocket reaches a height h.

[**Ans:** $T = m\left(f + \dfrac{R^2}{x^2}g\right)$; $W = mh\left(f + \dfrac{R}{R+h}g\right)$

where R is Earth's radius]

3. The potential energy function for the force between two atoms in a diatomic molecule can be expressed approximately as $\dfrac{a}{x^{12}} - \dfrac{b}{x^6}$. Find the force and the dissociation energy. [*Hint:* The disassociation energy equals the change in potential energy from the minimum value to the value at $x = \infty$].

[**Ans:** $\dfrac{12a}{x^{13}} - \dfrac{6b}{x^7}$; $\dfrac{b^2}{4a}$]

4. A bullet of mass m is fired with speed v into a cubical block of edge a and mass M. The block is suspended by strings so that its centre is a distance l from the axis of support. Determine the angle at which the block swings if it swings without rotation.

$$\left[\textbf{Ans: } \cos^{-1}\left[1-\frac{m^2v^2}{2gl(m+M)^2}\right]\right]$$

5. A light rod of length l has a mass m attached to its end, forming a simple pendulum. It is inverted than released. What is its speed at the lowest point and what is the tension T in the suspension at that instant? The same pendulum is put in a horizontal position and released from the rest. At what angle from the vertical will the tension in the suspension equal the weight in magnitude? $\left[\textbf{Ans: } 2\sqrt{gl},\ 5mg;\ 71°\right]$

6. A shell of mass M kg is travelling at V m/sec when there is an internal explosion and the shell splits into two parts; one of mass m moves off at v m/sec at an angle θ to the original line of travel.

(i) Show that the direction of motion of the other part makes an angle φ with the original direction of travel, where

$$\tan\varphi = \frac{mv\sin\theta}{MV - mv\cos\theta}$$

(ii) Determine the overall change in kinetic energy. Show that the expression has the same sign as of the numerical values, and hence, state whether the change is a gain or a loss.

$$\left[\textbf{Ans: } \text{(i) } \sqrt{\frac{[M^2V^2 - 2m\,MvV\cos\theta + m^2v^2]}{(M-m)}},\right.$$

$$\left.\text{(ii) } -\frac{mM}{2(M-m)}[(V-v)^2 + 2vV(1-\cos\theta)];\ \text{Gain}\right]$$

7. A gun is mounted on a railway truck which is free to run without friction on a straight horizontal railway track. The gun and truck, together of mass M, are moving along the track with velocity u, when a shell of mass m (not included in M), is fired from the gun with muzzle velocity v relative to the gun. If the gun barrel and truck lie in the same vertical plane, and the former is inclined at an angle α to the direction in which the truck is moving, find the horizontal range of the shell.

$$\left[\textbf{Ans: } \frac{2v\sin\alpha}{g}\left[u + \frac{Mv\cos\alpha}{M+m}\right]\right]$$

8. Two stars of masses m_1 and m_2 deemed to be unaffected by all other objects in the universe, are initially at a great distance apart and moving with velocities u_1 and u_2. They come close enough to influence each other's motion and then separate to a great distance, their velocities being v_1 and v_2. Explain why one would expect the initial and final kinetic energies of the system to be same? On this assumption, prove that

$$\mathbf{I} \cdot (\mathbf{U} + \mathbf{V}) = 0,$$

where **U**, **V** are the velocity of m_2 relative to m_1 at the beginning and the end of the period of effective interaction, and **I** the impulse of the force which m_1 exerts on m_2.

9. Find the velocity acquired by a block of wood, of mass M kg, which is free to recoil when it is struck by a bullet of mass m kg, moving with velocity v m/sec in a direction passing through its centre of mass. If the bullet be embedded a meters, show that the resistance of the wood to the bullet supposed uniform is $\frac{Mm}{M+m} \cdot \frac{v^2}{2ga}$ kg wt., and that the time of penetration is $\frac{2a}{v}$ s during which time the block will move $\frac{ma}{M+m}$ meters.

10. A particle of mass m is in harmonic oscillation along the axis ox, the law of oscillation being $x = a\cos(\omega t - \alpha)$. Find the laws describing the variations of the kinetic energy K, the potential energy V and the total energy E of the moving particle as functions of the co-ordinate x; the value of the potential energy V for $x = 0$ is taken to be zero.

$$\left[\textbf{Ans:} \quad K = \frac{1}{2}m\omega^2(a^2 - x^2),\ V = \frac{1}{2}m\omega^2 x^2,\ E = \frac{1}{2}m\omega^2 a^2\right]$$

11. A satellite is moving with the constant angular velocity $\boldsymbol{\omega}$ in such a way that a certain point P of it has the constant velocity $\mathbf{v}$. Prove that the velocity $\mathbf{V}$ of a point Q of the body is given by

$$\mathbf{V} = \mathbf{v} + \boldsymbol{\omega} \times (\mathbf{r}_q - \mathbf{r}_p),$$

where $\mathbf{r}_p$, $\mathbf{r}_q$ are the position vectors of P and Q, respectively. Prove that

$$\mathbf{v} \times \boldsymbol{\omega} \cdot (\mathbf{r}_q - \mathbf{r}_p) = -\mathbf{v}^2$$

is a necessary condition that $\mathbf{V}$ is perpendicular to $\mathbf{v}$. Can this condition be satisfied if $\mathbf{v}$ is parallel to $\boldsymbol{\omega}$?

12. A spaceship S of mass M is moving with velocity $\mathbf{U}$ under the action of no external forces. In order to alter the direction of motion, an instantaneous rocket burn takes place, during which gas of mass m is ejected with velocity $\mathbf{U} + \mathbf{v}$, the direction of $\mathbf{v}$ being inclined at an angle α to the direction of $\mathbf{U}$. Through what angle is the direction of the spaceship deviated?

$$\left[\textbf{Ans:} \quad \tan^{-1}\left[\frac{mu\sin\alpha}{\{mu\cos\alpha - (M-m)U\}}\right]\right]$$

13. An airplane flies at constant speed u in still air. It flies from a point O due east to a point P, and then returns to O and a cross wind is blowing with constant speed v from a direction α south of west. Show that the total time taken to return to O is $\dfrac{2c(u^2 - v^2 \sin^2 \alpha)^{1/2}}{(u^2 - v^2)}$, where c = OP.

14. A man swims round a triangular course ABC of each side equal to d in a straight canal, with side AB on one bank and the point C on the other bank. The speed of the man in still waters is u. If the water is flowing with constant speed v, then show that time of covering the course either way is

$$\frac{d(u + \sqrt{4u^2 - 3v^2})}{u^2 - v^2}$$

15. Discuss the recoil of a gun of mass M when a bullet of mass m is fired from it.

2

Forces in Three Dimensions

2.1 INTRODUCTION

Forces and couples are three-dimensional vectors and follow the vector rule of addition. But it must be remembered that while, for a rigid body, force is a line vector, couple is a free vector. Therefore, for combining forces the parallelogram rule of addition is needed, but for couples the triangle rule is enough. For determining the statical effect of a force its line of action has also to be prescribed; thus, two parallel forces of equal magnitude are not in general statically equivalent. However, the statical effect of a couple acting on a rigid body is completely determined by its magnitude and direction. Two couples are equivalent if the forces constituting their arms lie in parallel planes and have equal moduli of their moments and the same sense of rotation. Thus, there is no one-to-one correspondence between couples and the pair of forces forming the couples; to each pair of forces there corresponds a unique couple, but not conversely.

The action of a force on a rigid body, in general, is to displace the body, i.e. move it parallel to itself and to rotate it, the latter action is determined by the moment of the force which we proceed to define.

2.2 MOMENT OF A FORCE

Moment of a force about a point O is a vector quantity defined as

$$\mathbf{M} = \mathbf{r} \times \mathbf{F}, \tag{2.1}$$

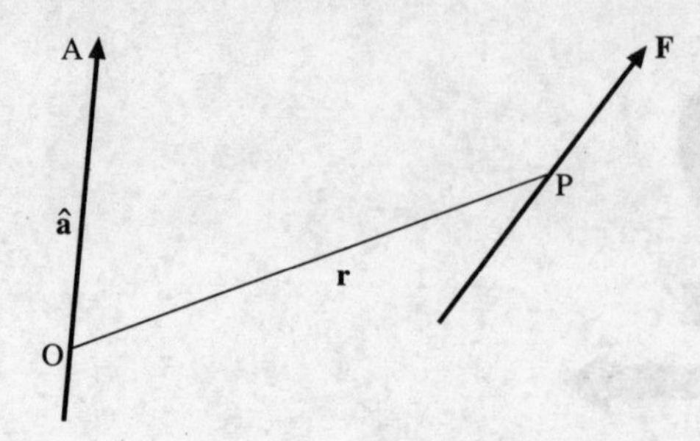

Figure 2.1 Axis $\hat{\mathbf{a}}$ for the moment of force F about the point O.

where $\mathbf{r}$ = OP is the position vector of a point on the line of action of $\mathbf{F}$. Thus, it may be understood as a vector, applied at a point O, whose magnitude is equal to the area of the parallelogram constructed on the vectors $\mathbf{r}$ and $\mathbf{F}$ and the direction along its normal. The moment of force about it is then a bound vector with point of application at O. But, according to the definition, the moment of a force is unchanged when we shift the point P on the line of action of the force.

Varignon's theorem

Moment due to a given force is the sum of moments due to its constituent forces

Now, if we have n forces $\mathbf{F}_s$ acting at n points $\mathbf{r}_s$ ($s = 1, 2, \ldots, n$) the total moment is

$$\mathbf{M} = \sum_{s=1}^{n} \mathbf{r}_s \times \mathbf{F}_s.$$

If, all forces pass through a single point $\mathbf{r}$, and we conveniently take each $\mathbf{r}_s = \mathbf{r}$, we have

$$\mathbf{M} = \mathbf{r} \times \mathbf{F} = \mathbf{r} \times \sum_{s=1}^{n} \mathbf{F}_S = \sum_{s=1}^{n} \mathbf{r} \times \mathbf{F}_s.$$

Moment of the force about a point on the axis is the projection on the axis of the moment of force about a point on the axis. Let us take a line OA parallel to unit vector $\hat{\mathbf{a}}$ (Figure 2.1). Taking dot product of both the sides of Eq. (2.1) with $\hat{\mathbf{a}}$, we get

$$\begin{aligned} M_a &= \text{Component of } M \text{ along } \hat{\mathbf{a}} = \mathbf{M} \cdot \hat{\mathbf{a}} \qquad (2.2) \\ &= (\mathbf{r} \times \mathbf{F}) \cdot \hat{\mathbf{a}} = \mathbf{r} \cdot (\mathbf{F} \times \hat{\mathbf{a}}) \\ &= \mathbf{r} \cdot \hat{\mathbf{n}}\, |\mathbf{F}| \sin\theta = dF_{\perp}. \end{aligned}$$

Here θ is the angle between the directions of $\mathbf{F}$ and $\hat{\mathbf{a}}$, and $\hat{\mathbf{n}}$ is unit vector perpendicular to both $\mathbf{F}$ and $\hat{\mathbf{a}}$ (given by the right handed screw), also $F_{\perp}$ is the component of the force $\mathbf{F}$ perpendicular to the direction of $\hat{\mathbf{a}}$, and d is the shortest distance between the line OA and the line of action of $\mathbf{F}$. $F_{\perp}$ is seen to lie in the plane perpendicular to OA. M_a is clearly independent of a position

O on OA and defines the moment of the force **F** about the axis OA. According to Eq. (2.2) it may be interpreted as the moment of the component of force **F** in a plane perpendicular to the line OA about the line OA. It may be seen that a set of forces that produces rotation, but no translation constitutes a couple.

2.3 THEOREMS ON THREE-DIMENSIONAL FORCE SYSTEM

In statics, we simplify the problem by reducing the given system of forces to an equivalent simple form. This is achieved through following theorems:

Theorem 2.3.1 A force **F** at a point A is equivalent to a force **F** through another point O plus a couple.

Proof: Let the force **F** act through the point A. At the other arbitrarily chosen point O, introduce two equal and opposite forces **F** and –**F** as shown in Figure 2.2; these three forces are clearly equivalent to single force **F** at A. Now the original force **F** at A, and the additional force –**F** at O form a couple of moment $\mathbf{G} = \mathbf{OA} \times \mathbf{F}$. Therefore, we conclude that the single force **F** at A is equivalent to a force **F** at O and the couple **G**.

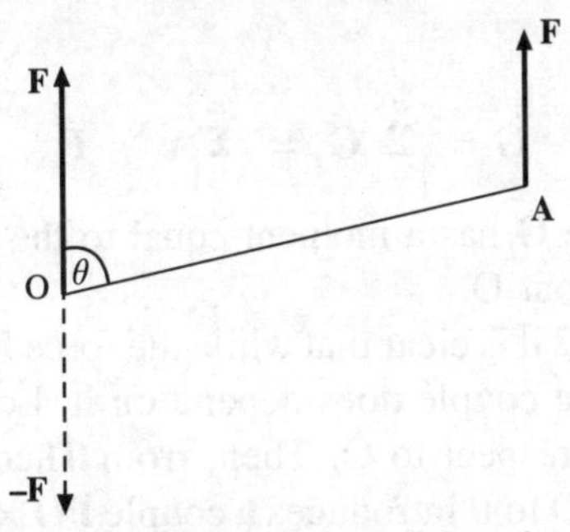

Figure 2.2 Shifting of force **F** from A to O.

Thus, we see that on shifting the force to a parallel position, a couple gets generated. **G** is a free vector of magnitude $|\mathbf{F}|d$ where d = OA sin θ is the perpendicular distance between O and line of action of **F** at A. Thus, since O is arbitrary, there is infinite number of arrangements possible to get the equivalent effects of the force. The process mentioned in this theorem is called reduction of the given force to the point O. The point O is known as the reduction centre and the couple $\mathbf{G} = \mathbf{r} \times \mathbf{F}$ as the associated or generated couple.

Theorem 2.3.2 Every force system is equivalent to a single force through an arbitrary point O and couple.

Proof: Let the force system (Figure 2.3) consist of n forces $\mathbf{F}_s$ acting at n points $\mathbf{A}_s$, where $s = 1, 2, \ldots, n$. It will be convenient to take O as the origin and assign $\mathbf{r}_s$ as the position vector of $\mathbf{A}_s$.

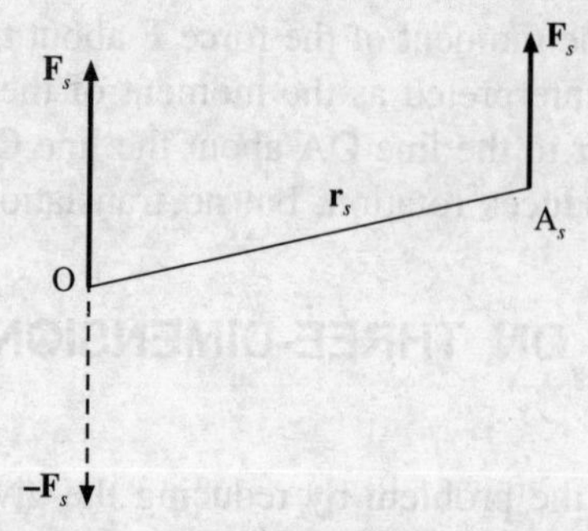

Figure 2.3 Shifting of the force system $\mathbf{F}_s$.

Now, from Theorem 2.3.1, we see that the force $\mathbf{F}_s$ at $\mathbf{A}_s$ is equivalent to a force $\mathbf{F}_s$ at O together with couples

$$\mathbf{G}_s = \mathbf{r}_s \times \mathbf{F}_s. \tag{2.3}$$

Since all forces act at O, and couples are free vectors, we conclude that the equivalent force system consists of a single force, which is given as

$$\mathbf{F} = \sum_{s=1}^{n} \mathbf{F}_s \text{ at O}, \tag{2.4}$$

and a couple

$$\mathbf{G} = \sum_{s=1}^{n} \mathbf{G}_s = \sum_{s=1}^{n} \mathbf{r}_s \times \mathbf{F}_s. \tag{2.5}$$

Observe that the couple **G** has a moment equal to the sum of moments of the individual forces $\mathbf{F}_s$ about O.

From Theorem 2.3.2 it is clear that while the force **F** does not depend on the choice of the origin, the couple does depend on it. Let P be the point having position vector **r** with respect to O. Then, from Theorem 2.3.1, we see that shifting the force from O to P introduces a couple $\mathbf{PO} \times \mathbf{F} = -\mathbf{OP} \times \mathbf{F} = -\mathbf{r} \times \mathbf{F}$. Thus, when the resultant force is taken at P, the total couple of the system is

$$\mathbf{G}' = \mathbf{G} - \mathbf{r} \times \mathbf{F}. \tag{2.6}$$

Equation (2.6) shows how the moment of the system changes with shift of the base point.

Theorem 2.3.3 If the force system has a single resultant then its moment about any point is equal to the sum of the moments of all the forces of the given system about the same point.

Proof: Suppose the force system $\mathbf{F}_s$ reduces to a single resultant **F** at a point P. Let O be an arbitrary point. Thus, with $\mathbf{G}' = 0$ and **G** given by Eq. (2.5), the result obtained in Eq. (2.6) shows

$$\mathbf{r} \times \mathbf{F} = \sum_{s=1}^{n} \mathbf{r}_s \times \mathbf{F}_s, \tag{2.7}$$

which proves the theorem. This theorem is known as Varignon's theorem.

Theorem 2.3.4 Every force system is equivalent to a single force and a couple whose moment is parallel to the single force.

Proof: From Theorem 2.3.2 we know that a force system can always be reduced to single force acting at a point O and a couple. Again at a point P (OP = **r**), the system is equivalent to the force **F** and a couple **G′** given by Eq. (2.6). Now, by the requirement of this Theorem, we seek a point **r** such that **G′** is parallel to **F** which on using Eq. (2.6) gives

$$\mathbf{G} - \mathbf{r} \times \mathbf{F} = p\mathbf{F}. \tag{2.8}$$

The above equations show that the moment of the couple and the force are parallel not only to a single point but to the set of points satisfying Eq. (2.8). It is easy to recognize Eq. (2.8) as the vector equation of a straight line.

Equation (2.8) may be put in the form

$$\mathbf{r} = \frac{\mathbf{F} \times \mathbf{G}}{|\mathbf{F}^2|} + t\mathbf{F}, \tag{2.9}$$

which is the vector equation of the straight line passing through the point $\dfrac{\mathbf{F} \times \mathbf{G}}{|\mathbf{F}^2|}$ and parallel to the direction **F**.

The reduction equation obtained in Eq. (2.9) is known as Poinsot's Reduction of a system of forces. The straight line represented by Eqs. (2.8) or (2.9) for a point along which the couple is parallel to the force is called Poinsot's Axis or Central Axis of the force system.

A force **F** together with a parallel couple **G** constitute a Wrench (**F**; **G**); thus, in this case

$$\mathbf{G} = p\mathbf{F}, \tag{2.10}$$

where p is some scalar. |**F**| and p are respectively termed *intensity and pitch of the wrench.* The action of a wrench is to produce a motion along a cork-screw path or helix. Hence, this special combination is also known as screw.

Poinsot's reduction shows that every force system can be put in the form of the wrench (**F**; **G′**) where **G′** is equal to $\mathbf{G} - \mathbf{r} \times \mathbf{F}$. Since the net force is not affected by shifting, it is an invariant of the system. Taking dot products of both sides of Eq. (2.8) with respect to **F**, we obtain

$$p = \frac{\mathbf{F} \cdot \mathbf{G}}{|\mathbf{F}^2|} \tag{2.11}$$

which on using Eq. (2.6) shows that the pitch p is also an invariant of the system, and so is $\Gamma = \mathbf{F} \cdot \mathbf{G}$. The pitch is positive if **F** and **G** in the wrench (**F**; **G**) have the same direction, i.e. if the wrench is right handed and similarly is negative if the wrench is left handed.

It is easy to see that when the force system reduces to a couple only, the pitch is infinite. However, the pitch becomes zero if it reduces to a single force.

Theorem 2.3.5 The necessary and sufficient condition that a given non-null force system reduces to a single force is that $\mathbf{F} \cdot \mathbf{G} = 0$.

Proof: The necessary part is obvious (?). For proving sufficiency we use the invariance property. Thus, we have for any position P where the couple is $\mathbf{G}'$

$$\mathbf{F} \cdot \mathbf{G}' = \mathbf{F} \cdot \mathbf{G} = 0. \tag{2.12}$$

The results obtained in Eqs. (2.11) and (2.12) show that either $\mathbf{G}' = 0$, which proves the theorem or that for all positions P, $\mathbf{G}'$ is perpendicular $\mathbf{F}$ as shown in Figure 2.4. Since $\mathbf{G}'$ is a free vector, we may consider it as composed of forces $\mathbf{F}$ and $-\mathbf{F}$, the former acts at the same point P where the force $\mathbf{F}$ acts and the latter at Q, such that PQ $|\mathbf{F}| = |\mathbf{G}'|$, with the line PQ at right angles to the force $\mathbf{F}$ and lying in the plane whose normal is along $\mathbf{G}'$. $\mathbf{F}$ and $-\mathbf{F}$ at P cancel out and we are left with a single force $\mathbf{F}$ at Q. This incidentally also proves that a coplanar force system (unless it is a couple) always reduces to a single force.

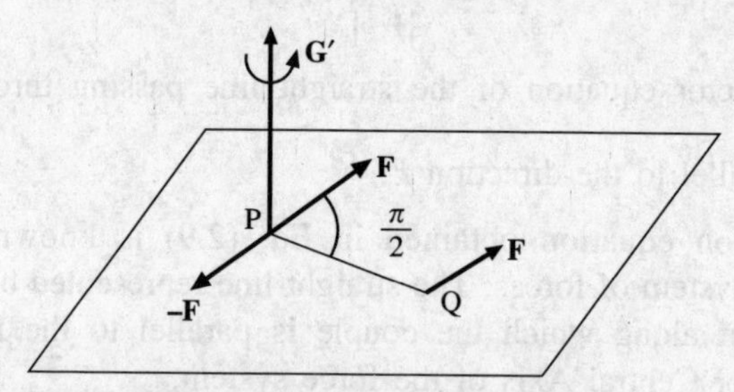

Figure 2.4 Couple $\mathbf{G}'$ equivalent to forces $\mathbf{F}$ (at Q) and $-\mathbf{F}$ (at P).

2.4 NULL LINES AND NULL PLANES

Let the system of forces be equivalent to a force $\mathbf{F}$ at O and a couple $\mathbf{G}$. Take any line perpendicular to $\mathbf{G}$ at O, and then the scalar moment of the system about this line is zero, because $\mathbf{F}$ meets the line and $\mathbf{G}$ is perpendicular to it. For this reason the line is called *null line* and the plane perpendicular to $\mathbf{G}$, generated by all such lines, is called *null plane* at the point O. Thus, we see that the single force in Theorem 2.3.5 lies in the null plane at P. If the point P lies in the null plane at O, then the line OP is a null line and so lies in the null plane at P. Thus, we see that if the null plane at O passes through P the null plane at P passes through O. This can also be seen from the equation to the null plane at the point $\mathbf{r}'$ which we now derive. From the result obtained in Eq. (2.6), we see that couple $\mathbf{G}'$ when the force is shifted to $\mathbf{r}'$, is given by

$$\mathbf{G}' = \mathbf{G} - \mathbf{r}' \times \mathbf{F}.$$

Hence, the equation to the plane through $\mathbf{r}'$ perpendicular $\mathbf{G}'$ is

$$(\mathbf{r} - \mathbf{r}') \cdot (\mathbf{G} - \mathbf{r}' \times \mathbf{F}) = \text{O},$$

or

$$(\mathbf{r} - \mathbf{r}') \cdot \mathbf{G} = (\mathbf{r} \times \mathbf{r}') \cdot \mathbf{F}. \tag{2.13}$$

This is the equation to the null plane at $\mathbf{r}'$. Observe the symmetry in $\mathbf{r}$ and $\mathbf{r}'$ and draw the conclusion—if the null plane at $\mathbf{r}'$ passes through $\mathbf{r}$ then the null plane at $\mathbf{r}$ passes through $\mathbf{r}'$.

Note: The students are advised to write down the Cartesian form of the null plane at (x', y', z').

Theorem 2.4.1 Any system of forces can be reduced to an equivalent system consisting of two forces, of which the line of action of one may be chosen arbitrarily. Such a pair of equivalent forces is called *conjugate forces* of the system.

Proof: Let the given force system be equivalent to a force $\mathbf{F}$ at an arbitrary point O and a couple $\mathbf{G}$

Therefore,

$$\mathbf{F} = \mathbf{F}_1 + \mathbf{F}_2, \tag{2.14}$$

where $\mathbf{F}_1$ is along a given direction $\hat{\mathbf{a}}$ and $\mathbf{F}_2$ is in a direction perpendicular to $\mathbf{G}$, i.e.

$$\mathbf{F}_1 = F_1\hat{\mathbf{a}} \text{ and } \mathbf{F}_2 \cdot \mathbf{G} = (\mathbf{F} - F_1\hat{\mathbf{a}}) \cdot \mathbf{G} = 0,$$

giving

$$\mathbf{F}_1 = \frac{\mathbf{F} \cdot \mathbf{G}}{|\mathbf{G} \cdot \hat{\mathbf{a}}|}\,\hat{\mathbf{a}}. \tag{2.15}$$

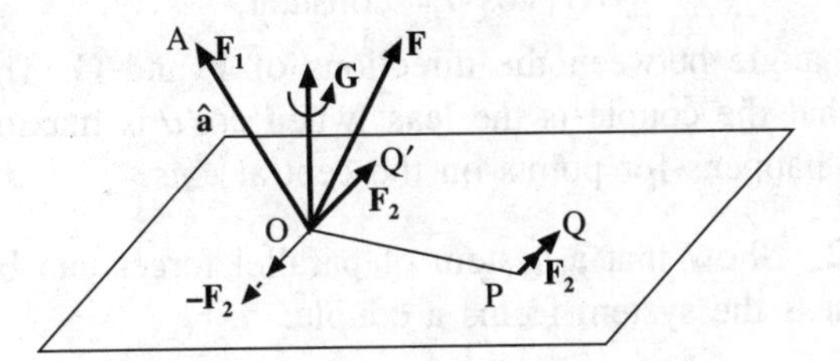

Figure 2.5 Conjugate forces $\mathbf{F}_1$ and $\mathbf{F}_2$.

Thus, $\mathbf{F}_1$ is seen to be known when choice has been made of O and $\hat{\mathbf{a}}$. By knowing the value of $\mathbf{F}_1$ the value of $\mathbf{F}_2 = \mathbf{F} - \mathbf{F}_1$, can be calculated.

Now $\mathbf{G}$ and $\mathbf{F}_2$ form a coplanar force system and so by Theorem 2.4.1 it can be reduced to a single force $\mathbf{F}_2$ acting along some definite line PQ in the null plane at O; this line is given by

$$\mathbf{G} = \mathbf{OP} \times \mathbf{F}_2. \tag{2.16}$$

$\mathbf{F}_1$ along OA and $\mathbf{F}_2$ along PQ constitute conjugate forces equivalent to the given system.

Theorem 2.4.2 The tetrahedron formed by conjugate forces is of constant volume.

Proof: Since O and $\hat{\mathbf{a}}$ are arbitrary it follows that there are infinite pairs of conjugate forces equivalent to the force system. Let $\mathbf{OA}$ $(= \mathbf{F}_1)$ and $\mathbf{PQ}$ $(= \mathbf{F}_2)$

represent a pair of conjugate forces equivalent to the force system (refer Figure 2.5). The four points O, A, P and Q may be taken to constitute a tetrahedron in space. From elementary knowledge of geometry it follows that the volume of the tetrahedron OAPQ is

$$V = \frac{1}{6}(\mathbf{PQ} \times \mathbf{OA}) \cdot \mathbf{OP} = \frac{1}{6}(\mathbf{F}_2 \times \mathbf{F}_1) \cdot \mathbf{r} \tag{2.17}$$

$$= \frac{1}{6}(\mathbf{r} \times \mathbf{F}_2) \cdot \mathbf{F}_1 = \frac{1}{6}\mathbf{G} \cdot (\mathbf{F} - \mathbf{F}_2)$$

$$= \frac{1}{6}\mathbf{F} \cdot \mathbf{G} = \frac{1}{6}p\,|\mathbf{F}^2|,$$

which is constant, since p and $|\mathbf{F}^2|$ are invariants.

SOLVED EXAMPLES

EXAMPLE 2.1 Show that, for any system of forces, the couple is least for points on the central axis.

Solution Since $|\mathbf{F}|$ and $\mathbf{F} \cdot \mathbf{G}$ are invariants, we have

$$|\mathbf{G}| \cos \theta = \text{constant},$$

where θ is the angle between the directions of **F** and **G**. The above result clearly shows that the couple is the least when $\cos \theta$ is maximum, i.e. when $\theta = 0$, and this happens for points on the central axis.

EXAMPLE 2.2 Show that a system of parallel forces can be reduced to a single force unless the system forms a couple.

Solution Let n parallel force be denoted by $\lambda_s \mathbf{f}$ and act at the position $\mathbf{r}_s$ $(s = 1, 2, \ldots, n)$. Suppose the force system reduces to a single force **F** at the origin O and a couple **G**. We have

$$\mathbf{F} = \sum_{s=1}^{n} \lambda_s \mathbf{f} = \lambda \mathbf{f}, \; \mathbf{G} = \sum_{s=1}^{n} \mathbf{r}_s \times \lambda_s \mathbf{f} = \mathbf{r}_c \times \mathbf{F},$$

where

$$\lambda = \sum_{s=1}^{n} \lambda_s \text{ and } \mathbf{r}_c = \sum_{s=1}^{n} \lambda_s \frac{\mathbf{r}_s}{\lambda}.$$

We easily see that if we shift the force **F** from 0 to the position $\mathbf{r}_c$, then the new couple is

$$\mathbf{G}' = \mathbf{G} - \mathbf{r}_c \times \mathbf{F} = 0.$$

Thus, except when $\lambda = 0$ (so that $\mathbf{F} = 0$) , i.e when the force system itself is a couple, it reduces to a single parallel force passing through the point $\mathbf{r}_c$, known as the **centre of force**.

From the formula giving its position, it is clear that the centre of force depends only on positions $\mathbf{r}_s$ and scalars λ_s and so remains unchanged by altering the directions of the forces. In the case of forces being weights, the centre is termed as centre of gravity.

EXAMPLE 2.3 If four forces are in equilibrium, show that the invariant $\Gamma = \mathbf{F} \cdot \mathbf{G}$ of any two forces is equal to the other two; also show that the same invariant of any three is zero.

Solution Let the four forces be $\mathbf{F}_1$, $\mathbf{F}_2$, $\mathbf{F}_3$, $\mathbf{F}_4$ at $\mathbf{r}_1$, $\mathbf{r}_2$, $\mathbf{r}_3$, $\mathbf{r}_4$, respectively. On transforming the forces to origin, following couples are introduced.

$$\mathbf{G}_1 = \mathbf{r}_1 \times \mathbf{F}_1,\ \mathbf{G}_2 = \mathbf{r}_2 \times \mathbf{F}_2,\ \mathbf{G}_3 = \mathbf{r}_3 \times \mathbf{F}_3,\ \mathbf{G}_4 = \mathbf{r}_4 \times \mathbf{F}_4. \quad \text{(i)}$$

Since the system in equilibrium, we must have

$$\mathbf{F}_1 + \mathbf{F}_2 + \mathbf{F}_3 + \mathbf{F}_4 = 0 \quad \text{(ii)}$$

and

$$\mathbf{G}_1 + \mathbf{G}_2 + \mathbf{G}_3 + \mathbf{G}_4 = 0. \quad \text{(iii)}$$

From Eqs. (ii) and (iii), we easily see that

$$(\mathbf{F}_1 + \mathbf{F}_2) \cdot (\mathbf{G}_1 + \mathbf{G}_2) = [-(\mathbf{F}_3 + \mathbf{F}_4)] \cdot [-(\mathbf{G}_3 + \mathbf{G}_4)] = [(\mathbf{F}_3 + \mathbf{F}_4)] \cdot (\mathbf{G}_3 + \mathbf{G}_4). \quad \text{(iv)}$$

Further, we have

$$(\mathbf{F}_1 + \mathbf{F}_2 + \mathbf{F}_3) \cdot (\mathbf{G}_1 + \mathbf{G}_2 + \mathbf{G}_3) = (-\mathbf{F}_4) \cdot (-\mathbf{G}_4) = \mathbf{F}_4 \cdot \mathbf{G}_4 = \mathbf{F}_4 \cdot (\mathbf{r}_4 \times \mathbf{F}_4) = 0 \quad \text{(v)}$$

the last term being written by using the property that the scalar triple product involving two parallel vectors vanishes identically. Since the suffix 1, 2, etc., have been chosen arbitrarily, the results obtained from Eqs. (iv) and (v) are seen to be true for all other such combinations.

EXAMPLE 2.4 Force act at the vertices of a tetrahedron outward, being perpendicular to the opposite faces and proportional to their areas; show that they are in equilibrium.

Solution Let OABC be the tetrahedron. With O as the origin supposes **a**, **b**, **c** are the position vectors of the points A, B, C, respectively.

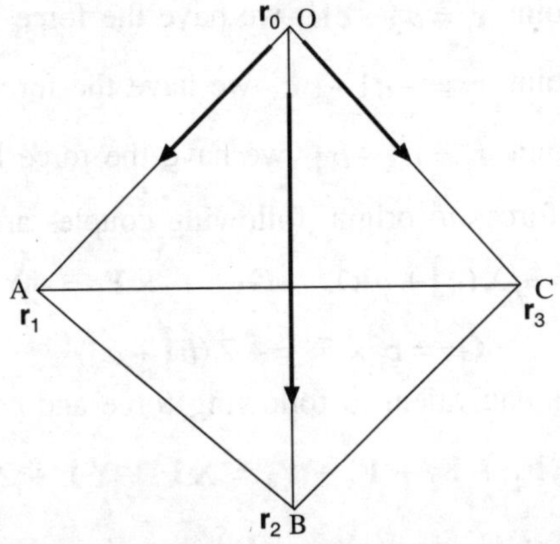

Figure 2.6 Force acting on the vertices of a tetrahedron.

From the conditions of the question, and keeping in view that the vector area of a triangle ABC is proportional to **AB** × **CA**, the given force system may be represented as

$$\text{At O; } \mathbf{r}_0 = O, \text{ force } \mathbf{F}_0 = \lambda\,[\mathbf{b} \times \mathbf{c} + \mathbf{c} \times \mathbf{a} + \mathbf{a} \times \mathbf{b}], \quad \text{(i)}$$

$$\text{At A, } \mathbf{r}_1 = \mathbf{a}, \text{ force } \mathbf{F}_1 = \lambda\,(\mathbf{c} \times \mathbf{b}), \quad \text{(ii)}$$

$$\text{At B, } \mathbf{r}_2 = \mathbf{b}, \text{ force } \mathbf{F}_2 = \lambda\,(\mathbf{a} \times \mathbf{c}), \quad \text{(iii)}$$

$$\text{At C, } \mathbf{r}_3 = \mathbf{c}, \text{ force } \mathbf{F}_3 = \lambda\,(\mathbf{b} \times \mathbf{a}). \quad \text{(iv)}$$

Transferring these forces to origin, following couples are generated:

$$\mathbf{G}_0 = \mathbf{r}_0 \times \mathbf{F}_0 = 0$$

$$\mathbf{G}_1 = \mathbf{r}_1 \times \mathbf{F}_1 = \lambda\,\mathbf{a} \times (\mathbf{c} \times \mathbf{b}) = \lambda\,[(\mathbf{a} \cdot \mathbf{b})\mathbf{c} - (\mathbf{a} \cdot \mathbf{c})\mathbf{b}],$$

$$\mathbf{G}_2 = \mathbf{r}_2 \times \mathbf{F}_2 = \lambda\,\mathbf{b} \times (\mathbf{a} \times \mathbf{c}) = \lambda\,[(\mathbf{b} \cdot \mathbf{c})\mathbf{a} - (\mathbf{b} \cdot \mathbf{a})\mathbf{c}],$$

$$\mathbf{G}_3 = \mathbf{r}_3 \times \mathbf{F}_3 = \lambda\,\mathbf{c} \times (\mathbf{b} \times \mathbf{a}) = \lambda\,[(\mathbf{c} \cdot \mathbf{a})\mathbf{b} - (\mathbf{c} \cdot \mathbf{b})\mathbf{a}].$$

Now, it is easy to see

$$\mathbf{F} = \text{Net force of the system} = \mathbf{F}_0 + \mathbf{F}_1 + \mathbf{F}_2 + \mathbf{F}_3 = 0,$$

$$\mathbf{G} = \text{Net couple of the system} = \mathbf{G}_0 + \mathbf{G}_1 + \mathbf{G}_2 + \mathbf{G}_3 = 0;$$

hence, the system is in equilibrium.

EXAMPLE 2.5 Forces X, Y and Z act along the three straight lines $y = b$, $z = -c$; $z = c$, $x = -a$; and $x = a$, $y = -b$, show that they will have a single resultant if $a\text{YZ} + b\text{ZX} + c\text{XY} = 0$, and that equations of its lines of action are any two of the three

$$\text{X}(y\text{Z} - z\text{Y}) - a\text{YZ} = 0,\ \text{Y}(z\text{X} - x\text{Z}) - b\text{ZX} = 0,\ \text{Z}(x\text{Y} - y\text{X}) - c\text{XY} = 0.$$

Solution Equations to the given lines can be put in the form

$$\frac{x}{1} = \frac{y-b}{0} = \frac{z+c}{0}; \quad \frac{x+a}{0} = \frac{y}{1} = \frac{z-c}{0}; \quad \frac{x-a}{0} = \frac{y+b}{0} = \frac{z}{1}.$$

Thus, the given force system can be represented as follows:

At the point $\mathbf{r}_1 = b\,\hat{\mathbf{j}} - c\,\hat{\mathbf{k}}$, we have the force $\mathbf{F}_1 = \text{X}\mathbf{i}$,

At the point $\mathbf{r}_2 = -a\,\hat{\mathbf{i}} + c\,\hat{\mathbf{k}}$, we have the force $\mathbf{F}_2 = \text{Y}\mathbf{j}$,

At the point $\mathbf{r}_3 = a\,\hat{\mathbf{i}} - b\,\hat{\mathbf{j}}$, we have the force $\mathbf{F}_3 = \text{Z}\mathbf{k}$.

On transferring the forces to origin, following couples are generated

$$\mathbf{G}_1 = \mathbf{r}_1 \times \mathbf{F}_1 = -\text{X}(c\,\hat{\mathbf{j}} + b\,\hat{\mathbf{k}}); \quad \mathbf{G}_2 = \mathbf{r}_2 \times \mathbf{F}_2 = -\text{Y}(a\,\hat{\mathbf{k}} + c\,\hat{\mathbf{i}});$$

$$\mathbf{G}_3 = \mathbf{r}_3 \times \mathbf{F}_3 = -\text{Z}(b\,\hat{\mathbf{i}} + a\,\hat{\mathbf{j}})$$

Hence, the system is equivalent to following force and couple at the origin

$$\mathbf{F} = \mathbf{F}_1 + \mathbf{F}_2 + \mathbf{F}_3 + \mathbf{F}_4 = \text{X}\,\hat{\mathbf{i}} + \text{Y}\,\hat{\mathbf{j}} + \text{Z}\,\hat{\mathbf{k}}, \quad \text{(i)}$$

$$\mathbf{G} = \mathbf{G}_1 + \mathbf{G}_2 + \mathbf{G}_3 + \mathbf{G}_4 = -[(c\text{Y} + b\text{Z})\,\hat{\mathbf{i}} + (a\text{Z} + c\text{X})\,\hat{\mathbf{j}} + (b\text{X} + a\text{Y})\,\hat{\mathbf{k}}]. \quad \text{(ii)}$$

Now, the necessary and sufficient conditions for the force system to reduce to a single force is $\mathbf{F} \cdot \mathbf{G} = 0$, which by the help of Eq. (i) and (ii) may be expressed as

$$[(cY + bZ)X + (aZ + cX)Y + (bX + aY)Z] = 0$$

or

$$aYZ + bZX + cXY = 0. \tag{iii}$$

Further when $\mathbf{F} \cdot \mathbf{G} = 0$, the force system reduces to a single force whose line of action is along the central axis having

$$\mathbf{G} - \mathbf{r} \times \mathbf{F} = 0, \tag{iv}$$

where

$$\mathbf{r} = x\hat{\mathbf{i}} + y\hat{\mathbf{j}} + z\hat{\mathbf{k}}$$

Thus, the component of Eq. (iv) are obtained as follows:

$$\begin{aligned} X(yZ - zY) - aYZ &= 0, \\ Y(zX - xZ) - bZX &= 0, \\ Z(xY - yX) - cXY &= 0, \end{aligned} \tag{v}$$

where Eq. (iii) has been used to establish the linear dependence of Eq. (v) of three equations; thus, only two of the three equations determine the line of action of the single resultant.

EXAMPLE 2.6 A Force **F** acts along the axis of x and another force $n\mathbf{F}$ along a generator of the cylinder $x^2 + y^2 = a^2$; show that the central axis lies on the cylinder.

Solution We are given two forces

$$\text{At } \mathbf{r}_1 = 0, \text{ force } \mathbf{F}_1 = \mathrm{F}\hat{\mathbf{i}},$$

$$\text{At } \mathbf{r}_2 = a\cos\theta\hat{\mathbf{i}} + a\sin\theta\hat{\mathbf{j}}, \text{ force } \mathbf{F}_2 = n\mathrm{F}\hat{\mathbf{k}}.$$

Transferring the forces to origin, following couples are generated:

$$\mathbf{G}_1 = \mathbf{r}_1 \times \mathbf{F}_1 = 0$$

$$\mathbf{G}_2 = \mathbf{r}_2 \times \mathbf{F}_2 = na\mathrm{F}(\sin\theta\hat{\mathbf{i}} - \cos\theta\hat{\mathbf{j}}).$$

Thus, we have

$$\mathbf{F} = \text{Total force of the system at O} = \mathbf{F}_1 + \mathbf{F}_2 = \mathbf{F}\hat{\mathbf{i}} + n\mathrm{F}\hat{\mathbf{k}}$$

Therefore,

$$\mathbf{r} \times \mathbf{F} = \mathrm{F}\,(ny\hat{\mathbf{i}} + (z - nx)\hat{\mathbf{j}} - y\hat{\mathbf{k}}].$$

Next, using above information, the vector equation

$$\mathbf{G} - \mathbf{r} \times \mathbf{F} = p\mathbf{F}$$

provides following component equations for the central axis

$$\frac{n\mathrm{F}\,(a\sin\theta - y)}{\mathrm{F}} = \frac{-\mathrm{F}\,(na\cos\theta + z - nx)}{0} = \frac{y\mathrm{F}}{n\mathrm{F}} = p$$

giving

$$n^2 a \sin\theta = y\,(1 + n^2),$$

$$na \cos\theta = nx - z.$$

Eliminating θ, the central axis is seen to lie on the cylinder

$$n^2(nx - z)^2 + (1 + n^2)y^2 = n^4 a^2.$$

EXAMPLE 2.7 Wrenches of the same pitch p act along the edges of a regular tetrahedron ABCD of side a. If the intensities of the wrenches along AB, DC are the same, and also those along BC, DA and DB, CA, show that the pitch of the equivalent wrench is $p + \dfrac{a}{2\sqrt{2}}$.

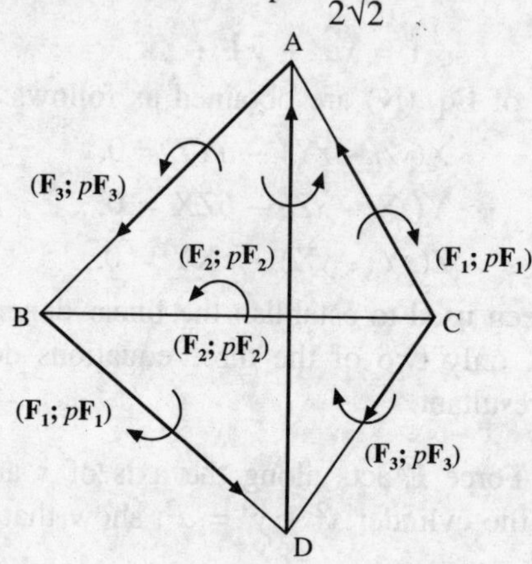

Figure 2.7 Tetrahedron ABCD formed of six wrenches.

Solution With D as the origin, let the position vectors of A,B,C be **a**, **b**, **c**. From the regularity of the tetrahedron, we have

$$|\mathbf{a}| = |\mathbf{b}| = |\mathbf{c}| = a, \tag{i}$$

$$\mathbf{a}\cdot\mathbf{b} = \mathbf{b}\cdot\mathbf{c} = \mathbf{c}\cdot\mathbf{a} = a^2 \cos\frac{\pi}{3} = \frac{a^2}{2}. \tag{ii}$$

The force system is given as

At $\mathbf{r} = 0$, forces $\mathbf{F}_1 = F_1\dfrac{\mathbf{a}}{a}$, $\mathbf{F}_2 = F_2\dfrac{\mathbf{b}}{a}$, $\mathbf{F}_3 = F_3\dfrac{\mathbf{c}}{a}$

and couples $\mathbf{G}_1 = p\mathbf{F}_1$, $\mathbf{G}_2 = p\mathbf{F}_2$, $\mathbf{G}_3 = p\mathbf{F}_3$;

At $\mathbf{r} = \mathbf{a}$, force $\mathbf{F}_1' = F_1\dfrac{\mathbf{b}-\mathbf{a}}{a}$; and couple $\mathbf{G}'_1 = p\mathbf{F}_1'$;

At $\mathbf{r} = \mathbf{b}$, force $\mathbf{F}_2' = F_2\dfrac{\mathbf{c}-\mathbf{b}}{a}$; and couple $\mathbf{G}'_2 = p\mathbf{F}_2'$; (iii)

At $\mathbf{r} = \mathbf{c}$, force $\mathbf{F}_3' = F_3\dfrac{\mathbf{a}-\mathbf{c}}{a}$; and couple $\mathbf{G}'_3 = p\mathbf{F}_3'$.

On transferring the force $\mathbf{F}'_1$, $\mathbf{F}'_2$, $\mathbf{F}'_3$ to origin D, following couples are introduced:

$$\mathbf{G}''_1 = \mathbf{a} \times \mathbf{F}'_1 = (\mathbf{a} \times \mathbf{b})\frac{F_1}{a},$$

$$\mathbf{G}''_2 = \mathbf{a} \times \mathbf{F}'_2 = (\mathbf{a} \times \mathbf{b})\frac{F_2}{a}, \qquad \text{(iv)}$$

$$\mathbf{G}''_3 = \mathbf{a} \times \mathbf{F}'_3 = (\mathbf{a} \times \mathbf{b})\frac{F_3}{a}.$$

Thus, at the origin, we have

$$\mathbf{F} = \text{Net force} = \mathbf{F}_1 + \mathbf{F}'_1 + \mathbf{F}_2 + \mathbf{F}'_2 + \mathbf{F}_3 + \mathbf{F}'_3$$

$$= \frac{[F_1(\mathbf{b}+\mathbf{c}-\mathbf{a}) + F_2(\mathbf{c}+\mathbf{a}-\mathbf{b}) + F_3(\mathbf{a}+\mathbf{b}-\mathbf{c})]^1}{a}, \qquad \text{(v)}$$

$$\mathbf{G} = \text{Net couple} = \mathbf{G}_1 + \mathbf{G}_2 + \mathbf{G}_3 + \mathbf{G}'_1 + \mathbf{G}'_2 + \mathbf{G}'_3 + \mathbf{G}''_1 + \mathbf{G}''_2 + \mathbf{G}''_3$$

$$= p\mathbf{F} + \frac{[F_1(\mathbf{a}\times\mathbf{b}) + F_2(\mathbf{b}\times\mathbf{c}) + F_3(\mathbf{c}\times\mathbf{a})]}{a}. \qquad \text{(vi)}$$

Now, we find that

$$|\mathbf{F}|^2 = 2(F_1^2 + F_2^2 + F_3^2), \qquad \text{(vii)}$$

and

$$\mathbf{F}\cdot\mathbf{G} = p\,|\mathbf{F}|^2 + \left(F_1^2 + F_2^2 + F_3^2\right)\frac{[\mathbf{a}\,\mathbf{b}\,\mathbf{c}]}{a^2} = |\mathbf{F}|^2\,\frac{p + [\mathbf{a}\,\mathbf{b}\,\mathbf{c}]}{2a^2}. \qquad \text{(viii)}$$

Here, $[\mathbf{a}\ \mathbf{b}\ \mathbf{c}]$ represents scalar triple product, and the cyclic property and the property of vanishing of this product, when two of the three vectors **a**, **b**, **c** are identical, have been used. Also, in deriving Eqs. (vii) and (viii), the values of scalar products given in Eq. (ii) have been employed.

From Eqs. (vii) and (viii), we get pitch of the equivalent wrench as

$$p' = p + \frac{\mathbf{F}\cdot\mathbf{G}}{|\mathbf{F}^2|} = p + \frac{3\,V}{a^2}, \qquad \text{(ix)}$$

where $[\mathbf{a}\ \mathbf{b}\ \mathbf{c}] = 6V$ is the volume of the tetrahedron ABCD. Substituting the value of V in Eq. (ix), we get the required result.

EXAMPLE 2.8 Two wrenches of pitches p, p' have axes at a distance $2a$ from one another. If the resultant wrench is of pitch q and its axis is equidistant from the axes of the component wrenches, show that the angle between them is

$$\tan^{-1}\frac{a(2q - p - p')}{a^2 - (q-p)(q-p')}.$$

Solution Let the two given wrenches (**F**; p**F**) and (**F**′; p' **F**′) act at A and B respectively where AB = $2a$ is the shortest distance line between the lines of action of the wrenches, and hence, is perpendicular to these two lines. The resultant wrench (**R**; q**R**) acts at O where OA = OB (See Figure 2.7).

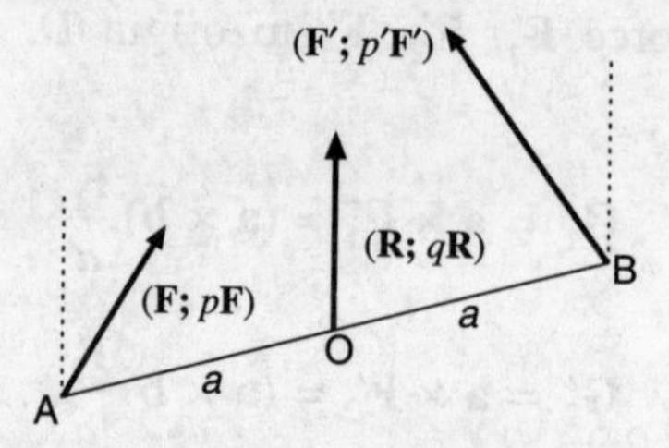

Figure 2.8 Resultant wrench (**R**; q**R**) of two wrenches (**F**; p**F**) and (**F′**; p'**F′**).

Now, from the equivalence of the force system

Net force:
$$\mathbf{R} = \mathbf{F} + \mathbf{F}', \tag{i}$$

Net couple:

$q\mathbf{R}$ = Given couple +
Couple generated in transferring the forces **F** and **F′** to O

$$= p\mathbf{F} + p'\mathbf{F}' + \mathbf{OA} \times \mathbf{F} + \mathbf{OB} \times \mathbf{F}'. \tag{ii}$$

Using Eq. (i), Eq. (ii) may be written as

$$(q - p)\mathbf{F} + (q - p')\mathbf{F}' = \mathbf{OA} \times \mathbf{F} + \mathbf{OB} \times \mathbf{F}'. \tag{iii}$$

Suppose θ is the angle between the lines of actions of the given wrenches. Taking dot product of both sides of Eq. (iii) with F, we get

$$(q - p)F^2 + (q - p')FF' \cos\theta = (\mathbf{OB} \times \mathbf{F}') \cdot \mathrm{F} = \mathbf{OB} \cdot (\mathbf{F}' \times \mathbf{F})$$
$$= \mathbf{OB} \cdot \hat{\mathbf{n}}\ FF' \sin\theta,$$

where $\hat{\mathbf{n}}$ is unit vector perpendicular to both **F′** and **F** and so is along **OB**. Therefore, the above equation simplifies to

$$(q - p)\,m + (q - p') \cos\theta = a \sin\theta, \tag{iv}$$

where
$$m = \frac{F}{F'}.$$

Similarly, taking dot product of both sides of Eq.(iii) with F′ and simplifying, we get

$$(q - p)m \cos\theta + (q - p') = m\,a \sin\theta. \tag{v}$$

Eliminating m between Eqs. (iv) and (v), we get the required result.

PROBLEMS

1. Derive the Cartesian equations of a central axis.
2. Show that pitch is an invariant of the force system. What are its dimensions?
3. Forces of magnitudes la, mb, nc act along three non-intersecting edges of a parallelepiped whose lengths are a, b, c, respectively. Prove that the invariant $\Gamma = \mathbf{F} \cdot \mathbf{G}$ of the system may be written as $(mn + nl + lm)V$, where V is the volume of the parallelepiped.

4. Show that any force system is in equilibrium if the resultant moments about three non-collinear points are all zero.
5. Show that a body cannot be in equilibrium under six forces acting along the edges of a tetrahedron.
6. An unknown force (X, Y, O) has a vector moment of (4, 1, –1) about the origin and acts on a body at the point (1, 3, 7). Find X and Y.

$$\left[\textbf{Ans:}\quad X = \frac{1}{7}, Y = \frac{-4}{7}\right]$$

7. Given two forces $\mathbf{F}_1$ and $\mathbf{F}_2$ acting on a body such that their turning effects about a point is the same. Does this mean that $\mathbf{F}_1 = \mathbf{F}_2$? If not, state the general relation between $\mathbf{F}_1$ and $\mathbf{F}_2$.

[**Ans:** $\mathbf{F}_2 = \mathbf{F}_1 + k(\mathbf{r}_2 - \mathbf{r}_1)$ for all k]

8. Does the following force system form a wrench? Force $\hat{\mathbf{j}} + \hat{\mathbf{k}}$ at (1, 1, 0), force at $2\hat{\mathbf{i}} - \hat{\mathbf{j}} - 3\hat{\mathbf{k}}$ at (2, 1, 0) and force $-\hat{\mathbf{i}} + 2\hat{\mathbf{k}}$ at (0, 0, 2).

[**Ans:** Yes]

9. Are the following forces in equilibrium $\mathbf{F}_1 = (3\hat{\mathbf{i}} - 2\hat{\mathbf{j}} - 4\hat{\mathbf{k}})$ at (1, 0, 1), $\mathbf{F}_2 = -\hat{\mathbf{i}} + \hat{\mathbf{j}}$ at (0, 1, 1), $\mathbf{F}_3 = -\hat{\mathbf{i}} + \hat{\mathbf{j}}$ at (0, 0, 1), $\mathbf{F}_4 = -\hat{\mathbf{i}} + 4\hat{\mathbf{j}}$ at (1, 1, 1).

[**Ans:** No]

10. A force with components (–7, 4, –5) acts at the point (2, 4, –3). Find its moment about the origin. Find also its moment about the line $x = y = z$, the positive sense of the line in which x increases.

[**Ans:** $(-8, 31, 36)$; $59\sqrt{3}$]

11. A rigid body is acted on by a force with components (1, 2, 3) at the point (3, 2, 1) and by a force with components (–1, –2, –3) at the point (–3, –2, –1). Find the equivalent force and couple at the origin.

[**Ans:** $\mathbf{F} = 0$, $\mathbf{G} = (8, -16, 8)$]

12. Forces represented by $F_0(2\hat{\mathbf{i}} - \hat{\mathbf{k}})$, $F_0(\hat{\mathbf{j}} + 4\hat{\mathbf{k}})$, $F_0(\hat{\mathbf{i}} - 3\hat{\mathbf{j}} + \hat{\mathbf{k}})$ – $F_0(\hat{\mathbf{j}} + \hat{\mathbf{k}})$ act on a rigid body, their points of application having the position vectors $a(2\hat{\mathbf{i}} + \hat{\mathbf{j}})$, $a\hat{\mathbf{j}}$, $2a\hat{\mathbf{j}}$, $a\hat{\mathbf{k}}$, respectively. Show that this system of forces is equivalent to a single force **F**. Determine **F** and the point in the xy-plane that is on its line of action.

$$\left[\textbf{Ans:}\ 3F_0(\hat{\mathbf{i}} - \hat{\mathbf{j}} + \hat{\mathbf{k}});\ a\left(-\frac{2}{3}\hat{\mathbf{i}} + 2\hat{\mathbf{j}}\right)\right]$$

13. The total moments of a system of forces about three points having the position vectors $a\hat{\mathbf{i}}$, $a(\hat{\mathbf{j}} + \hat{\mathbf{k}})$, $a(\hat{\mathbf{i}} + \hat{\mathbf{j}} + \hat{\mathbf{k}})$ are $F_0 a(-\hat{\mathbf{i}} + 2\hat{\mathbf{j}})$, $F_0 a(\hat{\mathbf{i}} + \hat{\mathbf{j}} + 3\hat{\mathbf{k}})$, $F_0 a(\hat{\mathbf{i}} + 2\hat{\mathbf{k}})$, respectively. Show that the system is equivalent to a single force. Find the components of the resultant force and the co-ordinates of a point on its line of action.

[**Ans:** $F_0(2\hat{\mathbf{i}} + \hat{\mathbf{j}} - \hat{\mathbf{k}})$; $a(3\hat{\mathbf{i}} + \hat{\mathbf{j}})$]

14. Calculate the turning moment about the origin of the force (3, 0, 0) P where line of action passes through (0, 1, 2). Suppose its line of action

had passed through (10, 1, 2); does this suggest a simplification of the three-dimensional problem?

[**Ans:** (0, 6, –3); no, both points lie on the line of action of the force]

15. A space station fires three of its rockets, located at the points (1, 1, 0), (1, 1, 1) and (1, 0, 1). If the forces exerted there are respectively $\hat{\mathbf{i}}+\hat{\mathbf{k}}$, $\hat{\mathbf{i}}+\hat{\mathbf{j}}$ and $-3\hat{\mathbf{j}}+\hat{\mathbf{k}}$; find their effect at the origin. Can this effect be achieved by a single rocket firing at some other point on the station?

[**Ans:** Force (2, –2, 2), couple (3, –1, –4); Yes]

16. A space station fires three of its rockets, located at the points (1, 1, 0), (0, –1, 1) and (1, 1, 1). If the forces exerted there are $3\hat{\mathbf{i}}+2\hat{\mathbf{k}}$, $4\hat{\mathbf{i}}+2\hat{\mathbf{j}}+\hat{\mathbf{k}}$ and $2\hat{\mathbf{i}}+\hat{\mathbf{j}}+\hat{\mathbf{k}}$, respectively; find their effect on the origin O. Can this effect be achieved by a single rocket firing along a single line? What are the effects of the above forces at the point (1, 1, 0) in the space station?

$$\left[\textbf{Ans: } \mathbf{F}=9\hat{\mathbf{i}}+3\hat{\mathbf{j}}+4\hat{\mathbf{k}},\ \mathbf{G}=-\hat{\mathbf{i}}+3\hat{\mathbf{j}};\ \text{yes, along } \frac{x}{9}=\frac{y}{3}=\frac{z-\frac{1}{3}}{4};\ \mathbf{G}'=-5\hat{\mathbf{i}}+7\hat{\mathbf{j}}+6\hat{\mathbf{k}}\right]$$

17. Two forces $\mathbf{F}_1$ and $\mathbf{F}_2$ act along non-intersecting lines. Prove that their central axis intersects the common perpendicular to the two lines, and divides it in the ratio $\mathbf{F}_2\cdot(\mathbf{F}_1+\mathbf{F}_2):(\mathbf{F}_1+\mathbf{F}_2)\cdot\mathbf{F}_1$.

18. Forces act along edges of a regular tetrahedron, viz. P along BC and DA, Q along CA and DB and R along AB and DC. Show that the pitch of the equivalent wrench is $\left(\frac{1}{(2\sqrt{2})}\right)$ of the edge of the tetrahedron.

19. Show that the minimum distance between two forces, which are equivalent to a given system (**F**; **G**) and which are inclined at a given angle 2α, is $\left(\frac{2\mathrm{G}}{\mathrm{F}}\right)\cot\alpha$, and that the forces are then each equal to $\left(\frac{1}{2}\right)\mathrm{F}\sec\alpha$.

20. Six equal forces act along the edges of a regular tetrahedron ABCD in the directions AB, BC, CA, DA, DB, DC. Prove that their central axis is perpendicular from D to the face ABC.

21. At every point of an octant of an ellipsoid cut off by the principal planes along the normal acts a force proportional to the element of surface at *P*. Show that these forces are equivalent to a single force acting along the line

$$a\left(x-\frac{4a}{3\pi}\right)=b\left(y-\frac{4b}{3\pi}\right)=c\left(z-\frac{4c}{3\pi}\right).$$

where $2a$, $2b$, $2c$ are the axes of the ellipsoid.

Method of Virtual Work

3.1 INTRODUCTION

Let us consider a body in equilibrium under a system of forces acting over it at various points. The forces may be real forces like gravity or forces of constraints like thrust in a rod. The problems in statics pertain to evaluating positions of equilibrium or determining constraint forces to preserve the equilibrium. Such problems may be solved by applying the equilibrium analysis given in Chapter 2. The method of virtual work provides an alternative route to achieve the goal that in many situations is easier to work out. For mathematical analysis, it is suffices to ignore the body and consider only the system of forces to be in equilibrium. In this form the system may be given some displacement, of course subject to certain inbuilt geometrical constraints. Virtual work is the work resulting from forces acting through a virtual displacement. In this discussion, the term displacement may refer to a translation or a rotation, and the term force not only refer to a force but also to a couple. The virtual displacements being independent variables are also arbitrary and this arbitrariness enables us to evaluate the desired unknown real quantities maintaining the equilibrium.

3.2 DISPLACEMENT

We know that the necessary and sufficient conditions for a system of n forces $\mathbf{F}_s$ ($s = 1, 2, ..., n$) acting at n points $\mathbf{r}_s$ to be in equilibrium are

(i) The vector sum of the forces is zero

$$\mathbf{F} = \sum_{s=1}^{n} \mathbf{F}_s = 0 \tag{3.1}$$

or in component form

$$F_x = F_y = F_z = 0; \tag{3.1a}$$

(ii) The vector sum of the moments of these forces about a point (say O) is zero

$$\mathbf{G} = \sum_{s=1}^{n} \mathbf{r}_s \times \mathbf{F}_s = 0 \tag{3.2}$$

or
$$G_x = G_y = G_z = 0. \tag{3.2a}$$

It may easily be seen that when the net force $\mathbf{F} = 0$ and the net moment $\mathbf{G} = 0$ for some point O, then the net moment will be zero for any other point. Also notice that $\mathbf{G} = \sum_{s=1}^{n} \mathbf{r}_s \times \mathbf{F}_s$ is the couple generated in the reduction of the force system to O.

The static equilibrium of a rigid body is determined by Eqs. (3.1) and (3.2), and the reader is already familiar with such problems in plane statics. We shall present here an alternate treatment known as the method of virtual work. But before doing so we shall study the displacements of a rigid body in space. Displacements are of two types:

Translation

A translation is a displacement in which each particle of the body receives the same vector displacement. An arbitrary point **r**, under the displacement ξ moves to $\mathbf{r} + \xi$. Here ξ is same for each point of the body, therefore it moves parallel to itself.

Rotation

A rotation is about a line or an axis; all the particles on this axis remain fixed. A rotation may be described by a vector along this line and having a magnitude given by the angle through which the body is turned (called *angular displacement*, the positive sense of the turning being given by the right handed screw. A remarkable theorem by Euler says, "*The most general displacement of a rigid body with a fixed point is equivalent to a rotation about a line through that point*".

It may be borne in mind that no rotation is involved in the displacement of a single particle or a point, it merely undergoes a translation. However, referred to an origin, an angular displacement may be given to the particle but that may again be described in terms of translation. In other words, an angular displacement of a point is equivalent to an associated translation as can be seen in Figure 3.1. (We can reach from A to P either by undergoing an angular displacement θ or by translations along AM and MP).

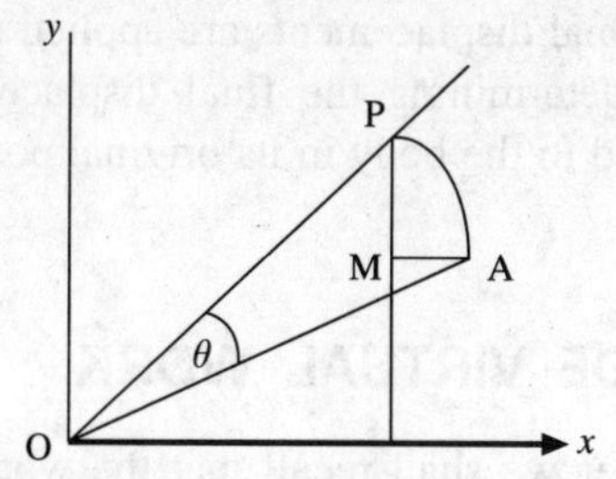

Figure 3.1 Translation associated with an angular displacement.

Thus, we can state that the most general displacement of a rigid body may be reduced to a translation followed by a rotation about some arbitrarily chosen base point or a rotation followed by translation. The rotation or angular displacement is same for all points of the body but the associated translation depends on the position of the point and this we calculate below in the case of an infinitesimal rotation.

Let a rigid body be turned through an infinitesimal angle $\delta\omega$ about an axis OA. On account of the rotation a point P of the body will trace arc PP′ of the curve with centre O′ lying on OA as shown in Figure 3.2. It is clear from the figure that

$$\text{Chord } PP' \simeq \text{arc } PP' = O'P\,\delta\omega = OP \sin\theta\,\delta\omega. \tag{3.3}$$

Moreover, it is seen that the direction of the small displacement $\mathbf{PP'}$ is perpendicular both to $\mathbf{OP}$ (= $\mathbf{r}$) and to the line $\mathbf{OA}$. Thus, keeping in view the direction and magnitude of $\mathbf{PP'}$ and exploiting the infinitesimal character of the angular displacement $\boldsymbol{\delta\omega} = \delta\omega\,\hat{\mathbf{a}}$, $\hat{\mathbf{a}}$ being unit vector along $\mathbf{OA}$, we have

$$\mathbf{PP'} = \boldsymbol{\delta\omega} \times \mathbf{r}, \tag{3.4}$$

corresponding to the small angular displacement $\delta\omega$.

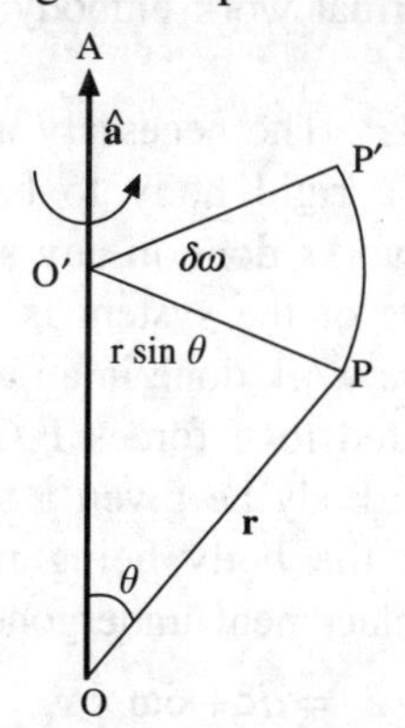

Figure 3.2 Angular displacement $\delta\omega$.

Further it is not difficult to see that the resultant of two infinitesimal rotations about the same point is the vector sum of these rotations. Also, the order in which infinitesimal displacements are applied to a rigid body does not matter. Therefore, for determining the final displacement, the infinitesimal displacements are applied to the body in its original position in any convenient manner and added.

3.3 PRINCIPLE OF VIRTUAL WORK

Before proceeding further we shall recall that the work done by a force $\mathbf{F}$ in a displacement $\mathbf{d}$ is $\mathbf{F} \cdot \mathbf{d}$.

When the displacement is not the actual but only deemed to be given, the displacement is said to be virtual and the ensuing work is called **virtual work**. Thus, by a virtual displacement of a system is understood every possible small, i.e. infinitesimal displacement that is compatible with the constraints imposed on the system. Constraints in mechanics imply the restrictions imposed on the displacement or position of the particles constituting the system. Only time independent geometric constraints are considered here and they are realized through strings, rods and surfaces.

Under a virtual displacement, a body gets displaced and deformed, and this results in virtual work done by the force system keeping it in equilibrium. When the body is rigid, as the case here is (with the exception of some problem concerning elastic strings), no work is done in deformation. Moreover, many-a-times, particularly in the case of a rigid body, force system can be taken as acting at a number of discrete points and we take the algebraic sum of the work done by forces separately. Moreover, the total work done by the forces in a number of displacements is the algebraic sum of the works done in the separate displacements. The use of the word algebraic stresses the fact that the appropriate +ve or –ve sign associated with the work must be taken into account while forming the sum. Now, we are in a position to enunciate and prove the principle of virtual work embodying the general conditions of equilibrium of a rigid body

Principle of Virtual Work: The necessary and sufficient condition for a system of forces acting on a rigid body to be in equilibrium is that the algebraic sum of the virtual works done in any small displacement consistent with the geometric constraints of the system is zero.

We shall first calculate the work done in an arbitrary virtual displacement. Let the rigid body be subjected to n forces $\mathbf{F}_s(X_s, Y_s, Z_s)$ $(s = 1, 2, ..., n)$ acting at the point $\mathbf{r}_s$. Let the body be given a translation $\delta\xi$, and a rotation $\delta\boldsymbol{\omega}$ about an arbitrary axis; the body being rigid these are same for all positions. Then the total displacement undergone by the position $\mathbf{r}_s$ is

$$\mathbf{d}_s = \delta\xi + \delta\boldsymbol{\omega} \times \mathbf{r}_s, \tag{3.5}$$

and so the work done by the force $\mathbf{F}_s$ is

$$\begin{aligned} \delta W_s = \mathbf{F}_s \cdot \mathbf{d}_s &= \mathbf{F}_s \cdot \delta\xi + \mathbf{F}_s \cdot (\delta\boldsymbol{\omega} \times \mathbf{r}_s) \\ &= \mathbf{F}_s \cdot \delta\xi + (\mathbf{r}_s \times \mathbf{F}_s) \cdot \delta\boldsymbol{\omega}, \end{aligned} \tag{3.6}$$

where we have used the cyclic property of scalar triple product. Hence, the net virtual work done by the forces of the system is

$$\delta W = \sum_{s=1}^{n} \delta W_s = \left(\sum_{s=1}^{n} \mathbf{F}_s \right) \cdot \delta \xi + \left(\sum_{s=1}^{n} \mathbf{r}_s \times \mathbf{F}_s \right) \cdot \delta \boldsymbol{\omega} \tag{3.7}$$

as it is permissible (?) to shift $\delta\xi$ and $\delta\boldsymbol{\omega}$ outside the summation sign. Now, identifying $\left(\sum_{s=1}^{n} \mathbf{F}_s \right) = \mathbf{F}$ as the net force and $\left(\sum_{s=1}^{n} \mathbf{r}_s \times \mathbf{F}_s \right) = \mathbf{G}$ as the net couple on the system, we have

$$\delta W = \mathbf{F} \cdot \delta \xi + \mathbf{G} \cdot \delta \boldsymbol{\omega}. \tag{3.8}$$

In proving the necessary part, we have to show that if the system is in equilibrium, then $\delta W = 0$. This is obvious because equilibrium implies $\mathbf{F} = \mathbf{G} = 0$.

In proving the sufficiency, we have to show that if $\delta W = 0$ for any displacement then the system is in equilibrium which again easily follows because of the arbitrariness of $\delta\xi$ and $\delta\boldsymbol{\omega}$.

The result

$$\delta W = 0 \tag{3.9}$$

is known as the *equation of virtual work.*

It should be remarked here that the principle of virtual work is actually a theorem providing an alternative method of handling equilibrium problems. Further, it is easy to see that the result holds (?) for coplanar forces too.

Forces acting on a system may be classified into two categories as follows:

(i) External forces such as the gravitational force appearing as the weight of a body

(ii) Internal forces of constraints, e.g., stress (tension or thrust) in a rigid rod forming a part of a framework or the reaction at a joint. These are reaction forces generated because the system is constrained to assume certain configuration.

In the principle of virtual work, all the forces doing work, whether external or internal, are to be taken into account while deriving the equation of virtual work. But in a free body diagram, the internal forces may also be marked as external forces.

Also, it is seen that, since the work is taken to be independent of the path, the principle holds only for conservative force system.

3.4 FORCES TO BE OMITTED

The advantage of using the method of virtual work is that we can prove that many types of forces perform no work while undergoing displacement, and so the equation of virtual work, viz. $\delta W = 0$, may be written down without much labour. Following are the five situations in which no work is done:

(i) The forces of mutual action and reaction between two points which remain a fixed distance apart.
Let A, B be the positions of two particles before displacement and A′, B′ their positions after displacement (Figure 3.3).

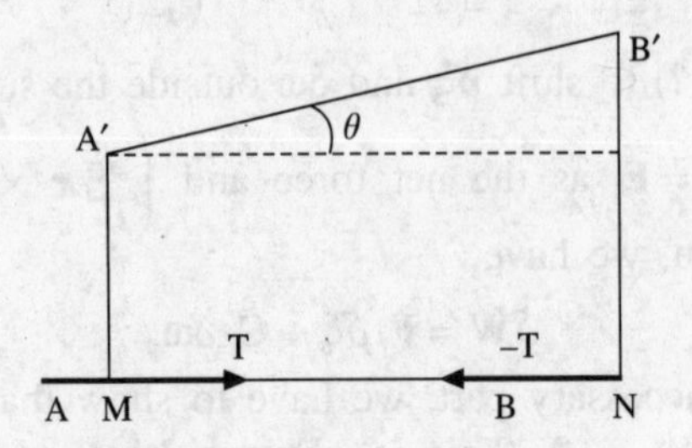

Figure 3.3 Work done by tension.

Suppose M and N are the projections of A′, B′, respectively on the line AB. The forces of action and reaction at A and B are T and –T as marked in the figure. It follows from definition that the work done by the pair of forces is given by

$$\delta W = T AM + (-T) BN \tag{3.10}$$

$$= T(AB - MB - MN + MB)$$

$$= T(AB - A'B' \cos\theta).$$

Since $A'B' \approx AB$, the above becomes

$$\delta W = \mathbf{T} \cdot \mathbf{AB}(1 - \cos\theta)$$

$$= \mathbf{T} \cdot \mathbf{AB} O(\theta^2). \tag{3.11}$$

Finally, under the small displacement assumption θ is small and so neglecting $O(\theta^2)$ terms, we have

$$\delta W = 0. \tag{3.12}$$

(ii) The reaction on a movable body in a smooth contact with a fixed body. This is evident since reaction at a smooth surface is normal to the surface.

(iii) The force pair of mutual action and reaction at a point of contact. At any contact there are two equal and opposite forces involved, one acting on one body and the other on the other body; when the point of contact gets displaced, the work done by the pair of forces are clearly seen to be equal but of opposite sign, therefore, the net work is zero.

(iv) The reaction on a body rolling without sliding on a fixed body. Let us now study rolling contact. Since only first order theory is being considered, it will be sufficient to investigate the rolling of a rigid circle C on a fixed line L (refer Figure 3.4).

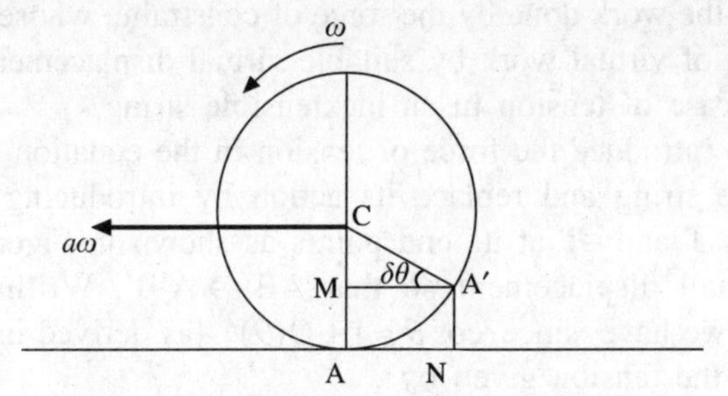

Figure 3.4 Rolling of a circle on a fixed line.

Let the circle of radius a role with angular speed ω so that as a whole of it is moving with horizontal velocity $a\omega$ towards the left. The initial point of contact A moves to A′, thereby covering an angle $\delta\theta = \omega\delta t$ in small time δt. But due to angular motion the point of contact will have an additional horizontal movement MA′ towards the right. Therefore, the displacement of the point of contact in small time δt is

Horizontally,

$$a\,\omega\delta t - \mathrm{MA}' = a\,\delta\theta - a\sin\delta\theta = O(\delta\theta)^3 \approx 0 \tag{3.13}$$

Vertically,

$$\mathrm{A'N} = a - a\cos\delta\theta = O(\delta\theta)^2 \approx 0.$$

Thus, we see that the point of contact remains momentarily at rest, and so the work done by the reaction is zero.

The fact that work is negligible at a rolling contact is of great importance in modern transport, and it also explains why machinery operates more efficiently on ball or roller bearings.

(v) The reaction at a fixed axis when a body is constrained to run round it. This is because the reaction always passes through the axis which itself is fixed.

It may be remarked here that the conclusion drawn above that no work is done by internal forces in small virtual displacement and translation is not surprising. It, embodying the principle of conservation of energy, reflects the idea that space is homogeneous. If we displace the system to a neighbouring position in space without disturbing it otherwise, the internal state of the system will remain unaffected, i.e. the distribution of various kinds of energy within the system remains unchanged, and hence, no work can be performed by the internal forces.

3.5 FORCES OF CONSTRAINTS

Many a times we need to find out the values of forces of constraints such as the tension in an inextensible string or the thrust in a rigid rod. This is done by introducing the work done by the force of constraint, whose value we need, in the equation of virtual work by suitable virtual displacement. We illustrate by taking the case of tension in an inextensible string.

In order to introduce the force of tension in the equation of virtual work, we remove the string and replace its action by introducing two equal and opposite forces T and –T at its end points as shown in Figure 3.3. Now we can give a small displacement so that AB → A′B′. Writing l = AB and $l + \delta l = \mathrm{A'B'}$, we have, since $\cos\theta \approx 1 + O(\theta)^2$ [as derived in Eq. (3.10)] the work done by the tension given by

$$\delta W = T\left[l - (l + \delta l)\cos\theta\right] \approx -T\delta l, \tag{3.14}$$

on neglecting square and higher powers of small quantities.

Similarly, since the thrust S in a rigid rod is in a sense opposite to that of tension, the work done by the thrust comes out as S δl.

SOLVED EXAMPLES

EXAMPLE 3.1 Four uniform rods are freely jointed at their extremities to form a parallelogram ABCD, which is suspended by the joint A, and is kept in shape by a string AC. Prove that the tension of the string is equal to half the whole weight.

Solution Let us replace the string AC by the two forces of constraint namely the forces of the tension T in the string as marked in Figure 3.5. Now, a displacement of $\mathrm{AC} \to \mathrm{AC} + \delta(\mathrm{AC})$ is given so that the length of the side rods do not alter. This will further ensure that no work is done by the stresses therein and by the reactions at the joints. The only other force doing work is the weight of the framework which can conveniently be taken at the centre of gravity G.

Taking the fixed point A as the origin from which we measure distances and calculate the displacements, we can write the equation of the virtual work as

$$-T\delta(\mathrm{AC}) + W\delta(\mathrm{AG}) = 0. \tag{i}$$

Using the geometrical fact that $\mathrm{AG} = \frac{1}{2}\mathrm{AC}$, Eq. (i) becomes

$$\left[-T + \frac{1}{2}W\right]\delta(\mathrm{AC}) = 0, \tag{ii}$$

which, because of the fact that although small, $\delta(\mathrm{AC}) \neq 0$, immediately gives the value of the tension as $T = \frac{1}{2}W$.

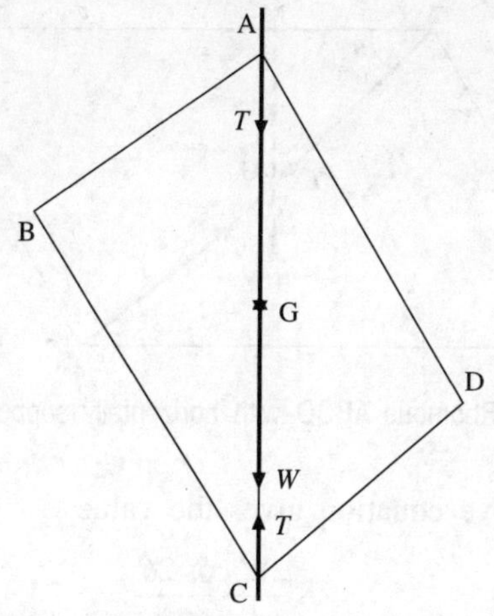

Figure 3.5 Parallelogram ABCD suspended from A.

[The above example illustrates all the points; the choice of suitable displacement and origin, the geometrical constraints and the omission of unwanted forces, inclusion of the work done by the sought of force of constraint and use of the geometry of the problem—needed to exploit the method of virtual work in solving equilibrium problems in statics].

EXAMPLE 3.2 A string, of length a forms the shorter diagonal of a rhombus formed of four uniform rods, each of length b and weight W, which are hinged together. If one of the rods be supported in a horizontal position, prove that the tension of the string is

$$\frac{2W(2b^2 - a^2)}{b\sqrt{(4b^2 - a^2)}}.$$

Solution Let ABCD be the rhombus with the rod AB supported in a horizontal position. The string forming the shorter diagonal AC is replaced by force of constraint T as marked in Figure 3.6. Suppose $\angle DAC = \angle BAC = \theta$.

Let us give the displacement $\theta \to \theta + \delta\theta$. The only work done is by the tension force T and the weight $4W$ of the frame work which may be taken at the centre of gravity G of the framework as marked in the figure. Then, taking AB as the fixed reference line, the equation of virtual work is

$$-T\delta(\text{AC}) + 4W\delta(\text{GM}) = 0. \tag{i}$$

Now, from the geometry of Figure 3.6, we have

$$\text{AC} = 2\text{AG} = 2b\cos\theta, \tag{ii}$$

$$\text{GM} = \text{AG}\sin\theta = b\sin\theta\cos\theta. \tag{iii}$$

Using the values obtained from Eqs. (ii) and (iii) in Eq. (i) and taking the variation, we get

$$b(2T\sin\theta + 4W\cos 2\theta)\,\delta\theta = 0.$$

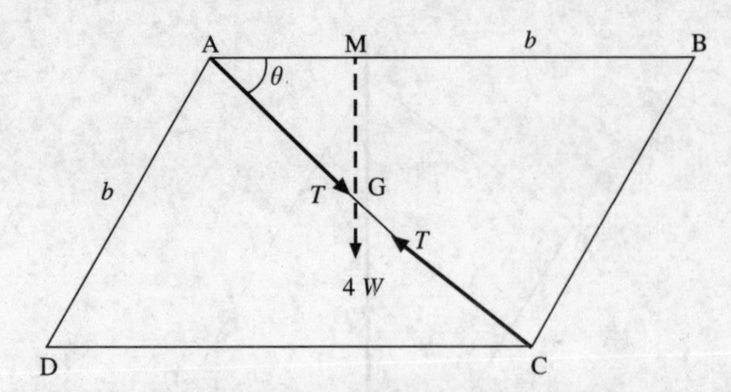

Figure 3.6 Rhombus ABCD with horizontally supported rod AB.

Since $\delta\theta \neq 0$, the above equation gives the value

$$T = -\frac{2W \cos 2\theta}{\sin \theta}. \qquad \text{(iv)}$$

Further, from the geometry of the problem, we know

$$\cos\theta = \frac{\text{AG}}{\text{AB}} = \frac{a}{2b}, \qquad \text{(v)}$$

because in equilibrium $\text{AG} = \frac{1}{2}\text{AC} = \frac{a}{2}$.

The student can now complete the question by substituting the values of $\cos 2\theta$ and $\sin\theta$ as obtained from Eq. (v) in (iv).

EXAMPLE 3.3 A heavy elastic string, whose natural length is $2\pi a$, is placed round a smooth cone whose axis is vertical and whose semi vertical angle is α. If W be the weight and λ the modulus of elasticity of the string, prove that it will be in equilibrium when in the form of a circle whose radius is

$$a\left(1 + \frac{W}{2\pi\lambda}\cot\alpha\right).$$

Solution Let us suppose that the radius of the circle in which the string lies is r. Give a displacement $r \to r + \delta r$. The forces doing work are the tension in the string and the weight of the string acting at its C.G. G at a depth OG = r cot a below the vertex O of the cone. Taking O as the fixed point, equation of virtual work is

$$-T\delta(2\pi r) + W\delta(r\cot\alpha) = 0,$$

or,

$$(2\pi T - W\cot a)\,\delta r = 0, \quad (\delta r \neq 0),$$

giving

$$T = \frac{W}{2\pi}\cot\alpha.$$

Use Hooke's law to write $T = \lambda\frac{r-a}{a}$, and complete the solution.

PROBLEMS

1. A smoothly jointed framework of light rods forms a quadrilateral ABCD. The middle points P, Q of an opposite pair of rods are connected by a string in a state of tension T, and the middle points R, S of the other pair by a light rod in a state of thrust X; show, by the method of virtual work, that $\frac{T}{PQ} = \frac{X}{RS}$. [*Hint:* Give a displacement such that $PQ \to PQ + \delta(PQ)$ and $RS \to RS + \delta(RS)$ without altering the lengths of the side rods. Then the equation of virtual work is $-T\delta(PQ) + X\delta(RS) = 0$. Next, proving geometrically that $AB^2 + CD^2 + 2PQ^2 = BC^2 + AD^2 + 2RS^2$ and get the relation $PQ\delta(PQ) = RS\delta(RS)$]
2. Show that the work done by a couple G in angular displacement $\delta\theta$ is $G\delta\theta$.
3. Work out the expression for δW in terms of translation components ($\delta\xi$, $\delta\eta$) and rotation $\delta\theta$ in two dimensions.
4. Six equal rods AB, BC, CD, DE, EF and FA are each of weight W and are freely joined at their extremities so as to form a hexagon; the rod AB is fixed in a horizontal position and the middle points of AB and DE are joined by a string, prove that its tension is $3W$.
5. A framework of four equal rods smoothly joined together lies in a vertical plane in the form of a square ABCD with the side AB clamped in a vertical position (B below A), and the middle points of AD and DC joined by means of a string. Find the tension in the string.

   [**Ans:** $4\sqrt{2}W$]
6. A frame consists of five bars forming the sides of a rhombus ABCD with the diagonal AC. If four equal forces P act inwards at the middle points of the sides, and at right angles to the respective sides, prove that the tension in AC is $\frac{P\cos 2\theta}{\sin\theta}$, where θ denotes the angle BAC.
7. Weights w_1, w_2 are fastened to a light inextensible string ABC at the points B, C the end A being fixed. Prove that, if a horizontal force P is applied at C and in equilibrium, AB, BC are inclined at angles θ and φ to the vertical, then $P = (w_1 + w_2)\tan\theta = w_2\tan\varphi$. [*Hint*: Use the fact that θ and φ are independent of each other].
8. A particle lies on a smooth horizontal table and is attached to three points, A, B, C forming an equilateral triangle of side $2a$ by three strings of natural length l, l' and l' of moduli λ, λ' and λ', respectively. Show that if the particle can rest in equilibrium at the centre O of the triangle, then $2a\left(\frac{\lambda}{l} - \frac{\lambda'}{l'}\right) = (\lambda - \lambda')\sqrt{3}$.

9. Two small smooth rings of equal weight slide on a fixed elliptic wire, whose major axis is vertical and they are connected by a string with passes over a small smooth peg at the upper focus; show that the weights will be in equilibrium wherever they are placed.

10. A uniform beam rests tangentially upon a smooth curve in a vertical plane and one end of the beam rests against a smooth vertical wall; if the beam is in equilibrium in any position, find the equation to the curve. [*Hint*: Take any horizontal line as the axis of x and the wall as the axis of y. The equation of virtual work will show that the height h of the C.G. of the rod is constant. The equation to the line representing the rod may be put in the form

$$y - h = \tan\theta(x - a\cos\theta)$$

where $2a$ is the length of the rod. Find the envelope of the above family of straight lines to get the required equation as

$$x^{\frac{2}{3}} + (y-h)^{\frac{2}{3}} = a^{\frac{2}{3}}$$

11. A parallelogram ABCD, of freely jointed rods, has the rod BC fixed, an inextensible string joins the middle points of AD, DC and a force P in the plane of the parallelogram is applied perpendicular to AB at its middle point; find the tension in the string when it is taut. [*Hint*: If θ is the angle ABC, then the work done by the force P is $\left(\frac{1}{2}\right)$ PAB$\delta\theta$ in the displacement $\theta \to \theta + \delta\theta$].

[**Ans:** $T = P\operatorname{cosec}\alpha$, where α is the angle which the string makes with CD]

12. A smooth cone of weight W stands inverted in a circular hole with its axis vertical. A string is wrapped twice round the cone just above the hole and pulled tight. What must be tension in the string so that it will just raise the cone?

$$\left[\textbf{Ans: } \left(\frac{W}{4\pi}\right)\cot\alpha\right]$$

13. An endless string of length $2\pi a + l (6a < 1 < 8a)$, passes around three equal smooth cylinders of weight W and radius a having their axes horizontal and parallel, and two of these rest on a horizontal plane, the third lying between them. Prove that the tension of the string is

$$\left(\frac{1}{2}\right)\frac{W(l-4a)}{[l(8a-l)]^{\frac{1}{2}}}.$$

14. Three equal and similar uniform rods, AB, BC, CD, freely jointed at B and C, have small smooth weightless rings attached to them at A and D. The rings slide on a smooth parabolic wire, whose axis is vertical and vertex upwards, and whose latus rectum is half the sum of the lengths of the three rods; prove that in the position of equilibrium, the inclination θ of AB or CD to the vertical is given by the equation

$$\cos\theta - \sin\theta + \sin 2\theta = 0.$$

15. A smooth rod passes through a smooth ring at the focus of an ellipse whose major axis is horizontal, and rests with its lower end on the quadrant of the curve which is furthest removed from the focus. Find its position of equilibrium, and show that its length must at least be $\frac{a}{4}\left[3+\sqrt{(1+8e^2)}\right]$, where $2a$ is the major axis and e is the eccentricity.

16. Three uniform rods OA, OB, OC, each of length a are freely jointed at a fixed point O, and have their upper extremities joined by uniform elastic strings each of natural length a. If the rods are equally inclined to the vertical and a sphere of radius l is placed between them, and if the weight of the sphere and the modulus of the strings are each equal to the weight of a rod, then if the strings are not in contact with the sphere, show that the inclination of the rods to the vertical is given by

$$2l \operatorname{cosec}^3\theta = 3a\left(6 - \sec\theta - 2\sqrt{3}\operatorname{cosec}\theta\right).$$

17. A prism whose cross-section is an equilateral triangle rests with two edges on smooth planes inclined at angles, α, β to the horizon. If θ be the angle which the plane containing these edges makes with the vertical show that

$$\tan\theta = \frac{2\sqrt{3}\sin\alpha\sin\beta + \sin(\alpha+\beta)}{\sqrt{3}\sin(\alpha-\beta)}.$$

18. A small heavy ring P slides on a smooth wire whose plane is vertical, and is connected by a string passing over a small pulley O in the plane of the curve with another weight W which hangs freely. If the ring is in equilibrium in any position on the wire, show that the form of the latter must be that of a conic section whose focus is at the pulley.

19. On a fixed circular wire (centre O and radius r) in a vertical plane slide two small smooth rings A and B, each of weight W. The rings are joined by light inextensible string of length $2a(< 2r)$ on which slides a small smooth ring C of weight P. Prove that for equilibrium either both parts of the string are vertical or else P is at a distance from the centre of the wire as follows

$$\left[\frac{W}{W+P}(r^2 - a^2)\right]^{\frac{1}{2}}.$$

20. Two uniform rods AB and AC, smoothly joined at A, are in equilibrium in a vertical plane, B and C rest on a smooth horizontal plane and the middle points of AB and AC are connected by a string. Show that the tension of the string is $\dfrac{W}{(\tan B + \tan C)}$.

21. A solid hemisphere is supported by a string fixed to a point as its rim and to a point on a wall with which the curved surface of the hemisphere is in contact. If θ and φ are the inclinations of the string and the plane base of the hemisphere to the vertical, prove that

$$\tan \varphi = \frac{3}{8} + \tan \theta.$$

22. A heavy rod of length $2l$, rests upon a fixed smooth peg at C with its end B upon a smooth curve if it rests in all positions, show that the curve is a conchord whose polar equation, with C as origin is

$$r = l + \frac{a}{\sin \theta}.$$

23. Three smooth equal circular cylinders, each of radius r, are placed inside a hollow cylinder of radius R, such that two of the smaller ones are in contact with the larger one and the third smaller one placed over the two. Show that equilibrium can exist only when $R < r\left(1 + \sqrt{28}\right)$.

24. ABCD is a parallelogram of freely joined rods; a point P on AB is joined to a point Q on CD by a string and a point R on AD to a point S on BC by another string. If the tensions in these strings be T and T' show that for equilibrium

$$\frac{T}{T'} = \frac{\text{PQ}}{\text{RS}} \cdot \frac{\text{AB}}{\text{AD}} \cdot \frac{\text{BS} - \text{AR}}{\text{AP} - \text{DQ}}.$$

25. A tripod is formed of three uniform rods, each of length $2a$ and weight w, freely joined together at the vertex. It stands on a smooth horizontal plane, and is prevented from collapsing by equal strings of length $a\sqrt{3}$ joining the feet. Prove that when a weight W is hung from the vertex the tension in each string is $\dfrac{w}{6} + \dfrac{W}{9}$.

Stability of Equilibrium

4.1 INTRODUCTION

Stability of a mechanical system is a very important quantity in physical problems. A system may be in equilibrium, yet it may fall apart even because of a small disturbance. Stability of a force system in equilibrium is determined by its response to a disturbance which in the first order linear theory, being investigated here, is taken to be small. Thus, let a system of forces in equilibrium be given a slight displacement, then, on the removal of the disturbing force in the displaced situation if the system tends to return to its original state, the equilibrium is said to be stable; if the system tends to move away from the original position of equilibrium, the equilibrium is said to be unstable. In the case, the displaced position is also a position of equilibrium, we have neutral equilibrium. The three situations are demonstrated by considering the case of a non-uniform sphere lying over a table (See Figure 4.1)

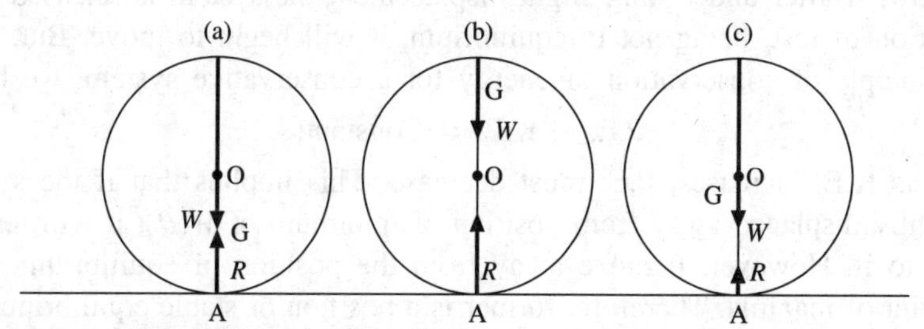

Figure 4.1 Three position of equilibrium (a) G below O (b) G above O (c) G to O.

Clearly the equilibrium conditions are

(i) $R = W$, where W is the weight of the sphere acting vertically downwards through the centre of gravity G and R is the reaction at the point of contact acting vertically upwards.

(ii) R and W are in the same vertical line, hence the centre O, the centre of gravity G and the point of contact A are in the same vertical line.

In the case of an off-centre sphere the equilibrium position shown in Figure 4.1(a), when the centre of gravity G is below the centre O, can easily be seen to be a position of stable equilibrium and the equilibrium position shown in Figure 4.1(b), when the centre of gravity G is above the centre O, a position of unstable equilibrium. Of course in Figure 4.1(c), when centre of gravity is at the centre, the equilibrium is neutral. The student is advised to argue out and arrive at the above conclusion. Cases of strictly neutral equilibrium are rare, being typified by Figure 4.1(c) of the sphere. Since there might be the possibility to impart such a small disturbance so as not to cause large deviation in an unstable position, it is necessary that the disturbance applied to an equilibrium state is arbitrary.

4.2 POSITION OF EQUILIBRIUM AND ITS STABILITY

Suppose the mechanical system has one degree of freedom designated by θ. Let the energy of the system in a state of rest be $W(\theta)$. Thus, $W(\theta)$ is the potential energy, i.e. the work done on the system by the external forces in bringing it to this state of rest from some other standard state of rest. Then, by its very definition, the work done in a small displacement is $\delta W = \left(\frac{dW}{d\theta}\right)\delta\theta$. From the principle of virtual work it now follows that this work must vanish in a position of equilibrium and conversely. Thus, the necessary and sufficient conditions for equilibrium are analytically expressible in terms of energy as

$$\frac{dW}{d\theta} = 0. \tag{4.1}$$

Now, if after undergoing slight displacement the system is released from a position of rest, being not in equilibrium, it will begin to move. But, from the principle of conservation of energy for a conservative system, we have

$$\text{P.E.} + \text{K.E.} = \text{Constant.} \tag{4.2}$$

Thus, as K.E. increases, P.E. must decrease. This implies that if the system is slightly displaced away from position of minimum of $W(\theta)$, it will tend to return to it. However, it move away from the position of equilibrium, were it a point of maxima. Therefore, former is a position of stable equilibrium and latter a position of unstable equilibrium. Thus, we make the following conclusion in the case of conservative forces.

Conclusion: Equilibrium occurs at positions of minima/maxima of the potential energy, the positions of minima being positions of stable equilibrium and positions of maxima positions of unstable equilibrium.

Analytically this may be expressed as

For positions of equilibrium, $W'(\theta) = \dfrac{\mathrm{d}W}{\mathrm{d}\theta} = 0.$

$$\text{For stable equilibrium,} \qquad W''(\theta) = \frac{\mathrm{d}^2W}{\mathrm{d}\theta^2} > 0. \tag{4.3}$$

$$\text{For unstable equilibrium,} \qquad W''(\theta) = \frac{\mathrm{d}^2W}{\mathrm{d}\theta^2} < 0.$$

The case for which $W''(\theta) = 0$ requires closer investigation; each particular problem is best treated according to its peculiarities. The above determination of stability of equilibrium is effectively identical to the investigation of the nature of the stationary points of the function $W(\theta)$. The cases in which $W''(\theta) = 0$ are the stability counter parts of those cases of stationary points of the function $W(\theta)$ for which some derivatives of higher order than the first also vanish. These cases often arise from the coincidence of a number of stationary points. What has been said about the system with one degree of freedom may be extended to a system with N degrees of freedom.

Here, we shall be mostly dealing with gravity as the external conservative force. Hence, the criterion for equilibrium and stability may be conveniently expressed in terms of the height z of the centre of gravity of the system. Thus, we have

$$\text{For positions of equilibrium,} \qquad \frac{\mathrm{d}z}{\mathrm{d}\theta} = 0,$$

$$\text{For stable equilibrium,} \qquad \frac{\mathrm{d}^2z}{\mathrm{d}\theta^2} > 0, \tag{4.4}$$

$$\text{For unstable equilibrium,} \qquad \frac{\mathrm{d}^2z}{\mathrm{d}\theta^2} < 0.$$

Thus, we see that a body is in stable position of equilibrium when the centre of gravity is in the lowest position it can take up, in which the centre of gravity is highest the body is in unstable equilibrium. In general, bodies which are top heavy are unstable and which are bottom heavy stable. This is how a tight rope walker balances himself by the help of a pole heavily weighted at one end.

4.3 BODIES OF CIRCULAR SECTION

Let a body with a circular section of radius r and centre O be placed over another fixed body of radius R and centre O_1. Suppose that the upper body can roll without sliding over the lower one. Suppose the initial point of contact is A and the common normal O_1AO vertical. After undergoing a small rolling displacement θ, so that the new point of contact is B, the displaced position of the centre O is O′ and the point A on the upper body assumes the position A′, represented by the angular displacement φ which is connected to θ by the relation

$$\text{Arc A}'\text{B} = \text{Arc AB}$$

or

$$r\varphi = R\theta. \tag{4.5}$$

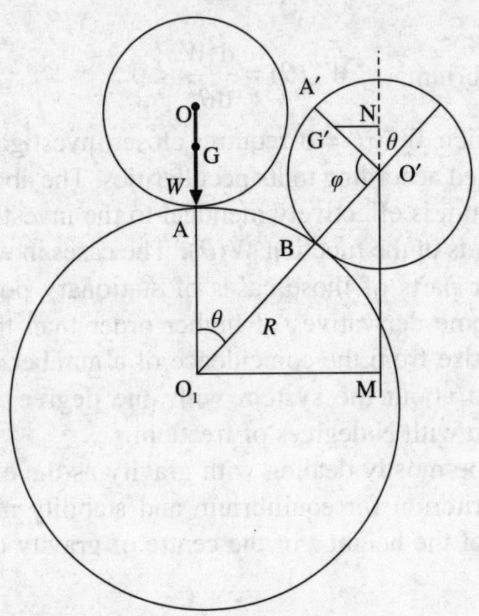

Figure 4.2 Equilibrium of a circular section body of radius *r* over another circular section body of radius *R*.

Let the height AG of the centre of gravity G above the point of contact A in the position of equilibrium be *h*. In the displaced position the centre of gravity is at G′, where A′G′ = AG = *h*, hence

$$\text{O}'\text{G}' = \text{OA}' - \text{A}'\text{G}' = r - h.$$

We can express the height z of G′ above the fixed position O_1 as

$$\begin{aligned} z &= \text{NM} = \text{NO}' + \text{O}'\text{M} = \text{O}'\text{M} + \text{NO}' \\ &= (R+r)\cos\theta + (r-h)\cos(\pi - \theta - \varphi) \\ &= (R+r)\cos\theta - (r-h)\cos\left(\frac{R+r}{r}\theta\right). \end{aligned} \tag{4.6}$$

Now, for equilibrium

$$\frac{\mathrm{d}z}{\mathrm{d}\theta} = -(R+r)\sin\theta + (r-h)\left(\frac{R+r}{r}\right)\sin\left(\frac{R+r}{r} - \theta\right). \tag{4.7}$$

Equation (4.7) is satisfied by $\theta = 0$, hence, the position in which the common normal is vertical is a position of equilibrium.

Next,

$$\left(\frac{d^2z}{d\theta^2}\right)_{\theta=0} = \left[-(R+r)\cos\theta + (r-h)\left(\frac{R+r}{r}\right)^2 \cos\left(\frac{R+r}{r}\theta\right)\right]_{\theta=0}. \quad (4.8)$$

The condition for stability viz.

$$\left[\frac{d^2z}{d\theta^2}\right]_{\theta=0} > 0$$

now provides

$$h < \frac{rR}{r+R} \text{ or } \frac{1}{h} > \frac{1}{r} + \frac{1}{R} \quad (4.9)$$

as the condition for the system to be in stable equilibrium. Of course, when

$$h > \frac{rR}{r+R} \text{ or } \frac{1}{h} < \frac{1}{r} + \frac{1}{R} \quad (4.10)$$

the equilibrium is unstable.

Remarks:

(i) When $\frac{1}{h} = \frac{1}{r} + \frac{1}{R}$, the determination of stability is much more complicated and the interested student is referred to Routh's Analytical Statics for further discussion.

(ii) If the upper body has a plane face in contact with the lower body then $r \to \infty$, hence the equilibrium is stable or unstable according as

$$h < \text{ or } > R. \quad (4.11)$$

(iii) If the lower body has a plane face then $R \to \infty$ and equilibrium is stable or unstable according as

$$h < \text{ or } > r. \quad (4.12)$$

(iv) Even when the sections near the point of contact are not circles, the same criterion can be used to investigate the stability, provided R and r are taken as the radii of curvatures of the lower and upper surface respectively at the point of contact.

(v) If the lower surface be concave near A, then replace R by $-R$. Thus, the equilibrium is stable or unstable according as

$$h < \text{ or } > \frac{rR}{R-r}. \quad (4.13)$$

(vi) If the common normal O_1AO is not vertical in the position of equilibrium but inclined at an angle α to the vertical, it can be shown that the equilibrium is stable or unstable according as

$$h < \text{ or } > \frac{rR}{R+r}\cos\alpha \quad (4.14)$$

where h again is the height of centre of gravity above the point of contact in the equilibrium position.

SOLVED EXAMPLES

EXAMPLE 4.1 A body consisting of a cone and a hemisphere rests on a rough horizontal table, the hemisphere being in contact with the table. Show that the greatest height of the cone, so that the equilibrium is stable, is $\sqrt{3}$ times the radius of the sphere.

Solution Let r be the radius of the hemisphere and x the height of the cone. A is the point of contact and G the centre of gravity of the body. Then, since lower body is plane, the equilibrium will be stable provided

$$r - \text{OG} = \text{AG} < r$$

or,

$$\text{OG} > 0$$

or,

$$V_1\text{OG}_1 - V_2\text{OG}_2 > 0,$$

where G_1 and G_2 are the positions of centre of gravity and V_1 and V_2, the volumes of the hemisphere and the cone respectively. Substituting the values

$$\text{OG}_1 = \frac{3}{8}r,\ \text{OG}_2 = \frac{1}{4}x,\ V_1 = \frac{2}{3}\pi r^3,\ V_2 = \frac{1}{3}\pi r^2 x \qquad \text{(i)}$$

in Eq. (i) and simplifying, we get

$$x < r\sqrt{3}$$

which proves the statement required to be proved.

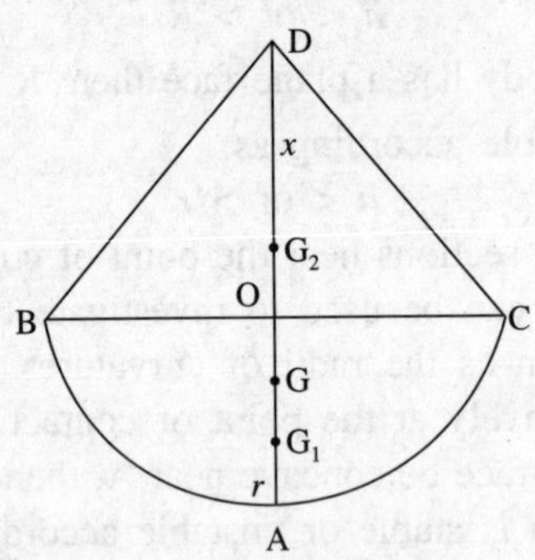

Figure 4.3 A body consisting of a cone of height x and hemisphere of radius r resting over a plane.

EXAMPLE 4.2 A uniform square lamina ABCD of side $2a$ rests in a vertical plane with sides AD and AB in contact with horizontal smooth pegs distant b apart, and in the same horizontal line. Find the positions of equilibrium and discuss their stability. Show in particular that, if $\frac{a}{\sqrt{2}} < b < a$, a non-symmetrical position of equilibrium is possible in which

$$\cos\theta = \frac{a}{b\sqrt{2}},$$

where θ is the inclination of the diagonal BD to the horizontal.

Solution Equilibrium is determined by the height z of the centre of gravity G above the fixed line PQ. We have

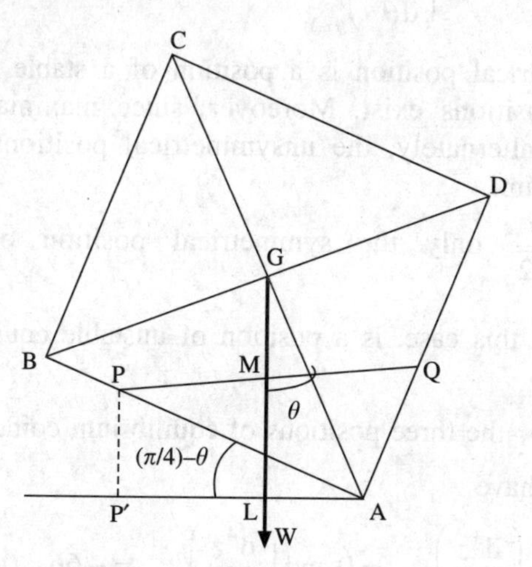

Figure 4.4 Square lamina ABCD of side 2*a* resting over smooth pegs at P and Q distance *b* apart.

$$z = \text{GM} = \text{GL} - \text{ML} = a\sqrt{2}\cos\theta - b\cos\left(\frac{\pi}{4}-\theta\right)\sin\left(\frac{\pi}{4}-\theta\right)$$

$$= a\sqrt{2}\cos\theta - \frac{b}{2}\cos 2\theta. \qquad \text{(i)}$$

This gives

$$\frac{dz}{d\theta} = -\left(a\sqrt{2} - 2b\cos\theta\right)\sin\theta \qquad \text{(ii)}$$

$$\frac{d^2z}{d\theta^2} = -\left(a\sqrt{2}\cos\theta - 2b\cos 2\theta\right). \qquad \text{(iii)}$$

Equilibrium positions occur when $\frac{dz}{d\theta} = 0$, i.e. when $\sin\theta = 0$, giving the symmetrical position $\theta = 0$, and when

$$\cos\theta = \frac{a}{b\sqrt{2}},$$

giving two unsymmetrical positions. These positions will exist only if such $\cos\theta$ exists $b < a$, i.e. if $\frac{a}{\sqrt{2}} < b$, and if the geometric condition $\theta < \frac{\pi}{4}$ is satisfied; this leads to the condition $\dfrac{a}{b\sqrt{2}} = \cos\theta > \dfrac{1}{\sqrt{2}}$ or to the condition $b < a$.

Now from Eq. (iii), when $a\sqrt{2} < 2b$, we see

$$\left(\frac{d^2 z}{d\theta^2}\right)_{\theta=0} = 2b - a\sqrt{2} > 0.$$

Thus, the symmetrical position is a position of a stable equilibrium, when unsymmetrical positions exist. Moreover, since maxima and minima are known to occur alternately, the unsymmetrical positions are positions of unstable equilibrium.

When $b < \dfrac{a}{\sqrt{2}}$ only the symmetrical position occurs and, since $\left(\dfrac{d^2 z}{d\theta^2}\right)_{\theta=0} < 0$ in this case, is a position of unstable equilibrium.

When $b = \dfrac{a}{\sqrt{2}}$, the three positions of equilibrium coincide in the position $\theta = 0$. Now, we have

$$\left(\frac{d^3 z}{d\theta^3}\right)_{\theta=0} = 0 \text{ and } \left(\frac{d^4 z}{d\theta^4}\right)_{\theta=0} = -6b < 0.$$

Whence we conclude that equilibrium in this case is unstable.

When $b > a$, we have only one position of equilibrium the symmetrical position $\theta = 0$; and, since $\left(\dfrac{d^2 z}{d\theta^2}\right)_{\theta=0} > 0$ in this case, it is a position of stable equilibrium.

(The students should reflect over the situation when $b > 2\sqrt{2}a$).

PROBLEMS

1. A hollow vessel made of thin uniform material consists of right circular cylinder of radius a and height h. One end of the vessel is closed by plane face and the other end is closed by a hemispherical shell. Show that the vessel can rest in the stable equilibrium with the hemisphere in contact with a horizontal table and the axis of symmetry of the vessel vertical provided $2h < \left(\sqrt{5} - 1\right)a$.
2. A cubical box of edge a is placed on the top of a fixed sphere, the centre of the force of the cube being in contact with the highest point of the sphere. Show that the least radius of the sphere for which the equilibrium will be stable is $\dfrac{a}{2}$.

3. A solid sphere rests inside a fixed rough hemispherical bowl of twice its radius. Show that, however large a weight is attached to the highest point of the sphere, the equilibrium is stable.

4. A solid hemisphere rests on a plane inclined to the horizon at an angle α. Find the greatest admissible value of α. If α be less than this value, is the equilibrium stable?

$$\left[\textbf{Ans:} \quad \sin^{-1}\left(\frac{3}{8}\right); \text{stable}\right]$$

5. A uniform smooth rod passes through a ring at the focus of a fixed parabola whose axis is vertical and vertex below the focus, and rests with one end on the parabola. Prove that the rod will be in equilibrium if it makes with the vertical an angle θ given by the equation

$$\cos^4 \frac{1}{2}\theta = \frac{a}{2c},$$

where $4a$ is the latus rectum and $2c$ the length of the rod. Investigate also the stability of the equilibrium in this position.

6. A uniform rod of length $2l$ has its lower end attached by a light string of length r to a point O, and it is constrained to pass through a fixed point A at a distance c vertically above O. Show that the rod is stable in the vertical position if

$$l < \frac{(c+r)^2}{r}.$$

7. One end A of a uniform rod AB of length $2a$ and weight w can turn freely about a fixed smooth hinge; the other end B is attached by a light elastic string of unstretched length a to a fixed support at the point O vertically above and distant $4a$ from A. If the equilibrium of the vertical position of the rod with B above A is stable, find the minimum modulus of elasticity of the string. [*Hint*: with $OAB = \theta$, the potential energy of the system is

$$W(\theta) = wa \cos\theta + \frac{1}{2}\lambda a\left[21 - 16\cos\theta - 4\sqrt{(5 - 4\cos\theta)}\right].$$

$$\left[\textbf{Ans:} \quad \left(\frac{W}{4}\right)\right]$$

8. Two rods, each of length $2a$, have their ends united at an angle α, and are placed in a vertical plane on a sphere of radius r. Prove that the equilibrium is stable or unstable according as $\sin\alpha >$ or $< \frac{2r}{a}$.

9. An isosceles triangular lamina of an angle 2α and height h rests between two smooth pegs at the same level, distant $2c$ apart; prove that if $3c \sec\alpha < h < 6c \operatorname{cosec} 2\alpha$, then oblique positions of equilibrium exists, which are unstable. Discuss the stability of the vertical position.

10. A heavy hemispherical shell of radius r has a particle attached to a point on the rim, and rests with the curved surface in contact with a rough

sphere of radius R at the highest point. Prove that if $\frac{R}{r} > \sqrt{5} - 1$, the equilibrium is stable, whatever be the weight of the particle.

11. A uniform hemisphere rests in equilibrium with its base upwards on the top of a sphere of double its radius. Show that the greatest weight which can be placed at the centre of the plane face without rendering the equilibrium unstable is one-eighth of the weight of the hemisphere.

12. A uniform plank of thickness $2h$ rests across the top of a fixed circular cylinder of radius a, whose axis is horizontal. Prove that the gain of potential energy when the plank is turned, without slipping, through an angle θ in a vertical plane parallel to its length, is $W[a\theta \sin \theta - (a + h)(1 - \cos \theta)]$ and deduce the condition of stability.

13. A heavy uniform rod rests partly within and partly without a fixed smooth hemispherical bowl, the rim of the bowl is horizontal and one point of the rod is in contact with the rim. Show that the equilibrium of the rod is stable.

14. A hemisphere rests in equilibrium on a fixed sphere of equal radius; show that the equilibrium is unstable when the curved, and stable when the flat surface of the hemisphere rests on the sphere.

15. Two equal particles are connected by a light string which is slung over the top of a smooth vertical circle; verify that the position of equilibrium is unstable. (It may be supposed that both particles rest on the circle, so that the length of the string is less than one-half of the circumference of the circle).

16. A uniform beam of length l rests with its ends on two smooth planes which intersect in a horizontal line. If the inclinations of the planes to the horizontal are α and β, β being the greater, show that the inclination θ of the beam to the horizontal, in one of the equilibrium positions, is given by $\tan \theta = \frac{1}{2}(\cot \alpha - \cot \beta)$ and show that the beam is unstable in this position.

17. A heavy uniform rod rests with one end against a smooth vertical wall and with a point on its length resting on a smooth peg. Find the position of equilibrium and show that it is unstable.

18. Show that a sphere partially immersed in a basin of water cannot rest in stable equilibrium on the summit of any convex part of the base.

19. A solid frustrum of a paraboloid of revolution, of height h and latus rectum $4a$, rests with its vertex on a paraboloid of revolution whose latus rectum is $4b$, show that the equilibrium is stable if

$$h < \frac{3ab}{a + b}.$$

20. A uniform elliptic cylinder of weight W is loaded with a particle of weight kW at one end of the major axis of the normal cross section through its centre of gravity, and is placed with its axis horizontal on a smooth horizontal table. Determine the possible positions of equilibrium and consider the stability of the symmetrical positions.

[**Ans:** When $e < \sqrt{\left\{\dfrac{k}{1+k}\right\}}$, two symmetrical positions of equilibrium one stable, the other unstable. When $e > \sqrt{\left\{\dfrac{k}{1+k}\right\}}$, four positions of equilibrium–the two symmetrical positions unstable.]

21. A heavy circular cylinder of height h rests on an inclined plane. If a be the radius of the base, show that the greatest inclination of the plane on which it can rest is $\tan^{-1}\left(\dfrac{2a}{h}\right)$.

22. A solid hemisphere rests in equilibrium on a rough inclined plane. Its spherical surface is in contact with the plane and its plane surface is vertical. If slipping is about to occur, then show that the coefficient of friction must be $\dfrac{3}{\sqrt{55}}$.

23. A cylinder of radius a whose axis OO′ is always horizontal can roll down a perfectly rough plane inclined at an angle α to the horizontal. The cylinder is eccentrically loaded so that its centre of gravity G is at a distance r from OO′. Show that if $r > a \sin \alpha$, equilibrium is possible for two positions of G relative to OO′, and that in each case the angle which the plane OO′G makes with the vertical is

$$\sin^{-1}\left(\frac{a \sin \alpha}{r}\right).$$

Also, show that only one of these positions is a position of stable equilibrium and find it.

24. A weight W is attached to an end A of a light rod AB of length l which is free to turn about B. A spring of unstretched length l is attached at its lower end to A and its upper end to a fixed point C which is vertically above B. The elastic properties of the spring are such that it doubles its natural length when extended by a weight W. Show that the system is stable when the rod is in a vertical position (i) if W is above B and $\text{BC} > \left(3+\sqrt{5}\right)\dfrac{l}{2}$, or (ii) if W is below B and $\text{BC} < \left(3+\sqrt{5}\right)\dfrac{l}{2}$.

[*Hint*: Energy stored in an elastic string is given by the product of mean tension and extension. [See Example 1.6 (Chapter 1)].

Equilibrium of Strings

5.1 INTRODUCTION

A perfectly flexible string or an ideal string is a thin material line which is inextensible and offers no resistance to bending and torsion. Therefore, the stress force exerted across any point of a perfectly flexible string is tension acting in the tangential direction, thereby simplifying the analysis considerably. Cables and chains used in many engineering structures can be regarded as perfectly flexible strings.

5.2 EQUATIONS OF EQUILIBRIUM

Let the string, in general a skew curve in a three dimensional space, be of line mass density m and subjected to external body force of force density $\mathbf{f}$ (force per unit mass) (see Figure 5.1). Consider the equilibrium of part AP, where A is a fixed position and P a variable position at an arc distance s from A. At the ends A and P act the tension forces $-\mathbf{T}_A$ and $\mathbf{T}$, respectively. These represent the action of the rest of the string on the part under consideration.

Now, the static equation of equilibrium can be easily written down in the integral form

$$\mathbf{T} - \mathbf{T}_A + \mathbf{F} = 0, \text{ where } \mathbf{F} = \int_A^P m\mathbf{f}\,\mathrm{d}s. \tag{5.1}$$

Differentiating this equation with respect to the arc distance s, we easily arrive at the differential equation of equilibrium

$$\frac{d\mathbf{T}}{ds} + m\mathbf{f} = 0. \tag{5.2}$$

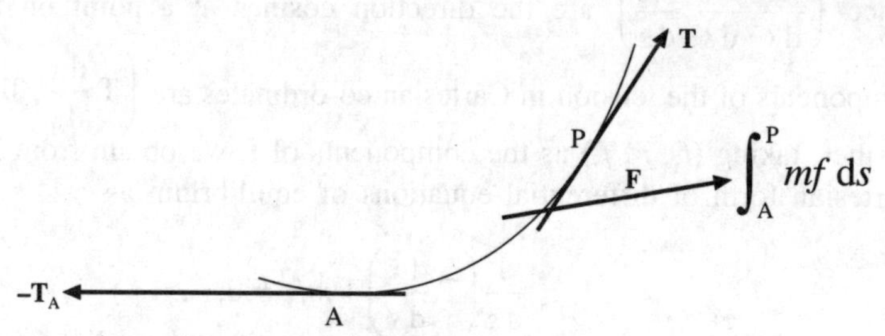

Figure 5.1 Force system on the string AP.

Since the tension is always along the tangent, we have

$$\mathbf{T} = \mathrm{T}\,\hat{\mathbf{t}}, \tag{5.3}$$

where $\hat{\mathbf{t}}$ is unit tangent and T is the magnitude of tension.

Using the well known result (see Frenet formulae) $\dfrac{d\hat{\mathbf{t}}}{ds} = \dfrac{\hat{\mathbf{n}}}{\rho}$, where $\hat{\mathbf{n}}$ is unit principal normal in the inward direction and ρ the radius of curvature, we obtain from Eq. (5.3)

$$\frac{d\mathbf{T}}{ds} = \frac{d\mathrm{T}}{ds}\hat{\mathbf{t}} + \frac{\mathrm{T}}{\rho}\hat{\mathbf{n}}. \tag{5.4}$$

Taking the natural trihedron with unit vectors $\hat{\mathbf{t}}, \hat{\mathbf{n}}, \hat{\mathbf{b}}$, ($\hat{\mathbf{n}}$ being unit principal normal and $\hat{\mathbf{b}}$ the binormal), we write

$$\mathbf{f} = \mathrm{f}_t\,\hat{\mathbf{t}} + \mathrm{f}_n\,\hat{\mathbf{n}} + \mathrm{f}_b\,\hat{\mathbf{b}}. \tag{5.5}$$

On substituting the value of $\dfrac{d\mathbf{T}}{ds}$ from Eq. (5.4) in Eq. (5.2) and equating the components in tangential and normal directions, we obtain the component differential equations of equilibrium as

Tangential: $$\frac{d\mathrm{T}}{ds} + m\,\mathrm{f}_t = 0, \tag{5.6}$$

Normal: $$\frac{\mathrm{T}}{\rho} + m\,\mathrm{f}_n = 0. \tag{5.7}$$

The third equation in the binormal direction gives $\mathrm{f}_b = 0$ indicating that at each point the string sets itself in the osculating plane there.

5.2.1 Cartesian Equations

Since $\left(\frac{dx}{ds}, \frac{dy}{ds}, \frac{dz}{ds}\right)$ are the direction cosines at a point on a curve, the components of the tension in Cartesian co-ordinates are $\left(T\frac{dx}{ds}, T\frac{dy}{ds}, T\frac{dz}{ds}\right)$. Further, taking (f_x, f_y, f_z) as the components of **f**, we obtain from Eq. (5.2) the Cartesian form of differential equations of equilibrium as

$$\frac{d}{ds}\left(T\frac{dx}{ds}\right) + mf_x = 0,$$

$$\frac{d}{ds}\left(T\frac{dy}{ds}\right) + mf_y = 0, \tag{5.8}$$

$$\frac{d}{ds}\left(T\frac{dz}{ds}\right) + mf_z = 0.$$

In two-dimensional situation the third equation is to be omitted.

The curve in which a string hangs freely under gravity can be shown (?) to be a plane curve lying on a vertical plane. One can find that in general, for a string hanging under gravity the horizontal component of tension has a constant value T_0, and that the differential equation of equilibrium satisfied by it has the form

$$\frac{d^2 y}{dx^2} = \frac{mg}{T_0}\frac{ds}{dx} = \frac{mg}{T_0}\sqrt{1+\left(\frac{dy}{dx}\right)^2}. \tag{5.9}$$

In the case of common catenary m = Constant.

5.3 COMMON CATENARY

When the external force system **f** is the force of gravity, the curve in which the string hangs is known as catenary and becomes a common catenary, if the line mass density is constant. Curve is easily seen to be a plane curve lying in a vertical plane.

Taking y-axis to be vertically upwards, we have

$$f_x = 0,\ f_y = -g, \tag{5.10}$$

so that

$$f_t = -g\sin\psi,\ f_n = -g\cos\psi \tag{5.11}$$

where ψ is the slope of the tangent (see Figure 5.2).

Substituting the values of f_t and f_n from Eq. (5.11) in Eqs. (5.6) and (5.7), we obtain

$$\frac{dT}{ds} = mg \sin \psi = mg \frac{dy}{ds} \tag{5.12}$$

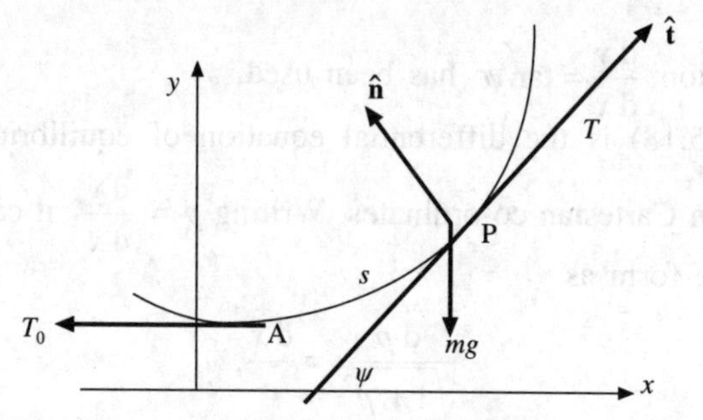

Figure 5.2 Force system for the common catenary with vertex at A.

and
$$T \frac{d\psi}{ds} = mg \cos \psi. \tag{5.13}$$

Notice that $\rho = \dfrac{ds}{d\psi}$ is the radius of curvature. From Eqs. (5.12) and (5.13), it follows that

$$\frac{dT}{T} = \tan\psi \, d\psi \tag{5.14}$$

Integrating Eqs. (5.13) and (5.14), we obtain

$$T \cos \psi = T_0 = w\ c, \text{ (say)}. \tag{5.15}$$

Equation (5.15) shows that horizontal component of tension is constant.

Substituting the value of T from Eq. (5.15) in Eq. (5.14), we get the value of radius of curvature as

$$\rho = \frac{ds}{d\psi} = c \sec^2 \psi. \tag{5.16}$$

Further, on integrating Eq. (5.16), the intrinsic equation of common catenary as

$$s = c \tan \psi. \tag{5.17}$$

Equation (5.17) can be derived directly by using the force system Eq. (5.10) in the equation of equilibrium Eq. (5.1) in the integral form and the student is advised to do so.

When there is no cause of confusion we shall find it convenient to delete the word common and call the curve represented by Eq. (5.17) or the curve in which a perfectly flexible inextensible string of uniform density hangs freely under the force of gravity *Catenary*.

5.3.1 Cartesian Equation of Catenary

Differentiating both sides of Eq. (5.17) w.r.t. ψ, we get

$$\frac{\mathrm{d}s}{\mathrm{d}x} = c\frac{\mathrm{d}^2 y}{\mathrm{d}x^2}, \tag{5.18}$$

where the relation $\dfrac{\mathrm{d}y}{\mathrm{d}x} = \tan\psi$ has been used.

Equation (5.18) is the differential equation of equilibrium of catenary [c.f. Eq.(5.9)] in Cartesian co-ordinates. Writing $p = \dfrac{\mathrm{d}y}{\mathrm{d}x}$, it can be expressed in the separable form as

$$\frac{\mathrm{d}p}{\sqrt{1+p^2}} = \frac{\mathrm{d}x}{c}.$$

Integrating, we get

$$\frac{\mathrm{d}y}{\mathrm{d}x} = p = \sinh\frac{x}{c}, \tag{5.19}$$

where the constant of integration vanishes provided $\psi = 0$ when $x = 0$, i.e., if we take the tangent to be horizontal at the point of intersection of the curve with y-axis. This point A, which is also the lowest point (?), is called the vertex of the catenary.

Integrating Eq. (5.19), we obtain

$$y = c\cosh\frac{x}{c}. \tag{5.20}$$

The constant of integration vanishing this time also, if we take $y = c$ when $x = 0$; this implies that the vertex is at a height c above the x-axis which is the directrix of the catenary. The parameter c governs the size of the catenary; the catenaries having the same value of c are identical except for their positions in space.

Now, using the result $\dfrac{\mathrm{d}y}{\mathrm{d}x} = \tan\psi$ in Eq. (5.19), we get

$$\sinh\frac{x}{c} = \tan\psi \tag{5.21}$$

and so

$$\cosh\frac{x}{c} = \sec\psi, \tag{5.22}$$

which combines with Eq. (5.20) to provide

$$y = c\sec\psi. \tag{5.23}$$

Now, adding Eq. (5.21) and (5.22), we obtain

$$x = c\log(\sec\psi + \tan\psi) = c\log\frac{y+s}{c}. \tag{5.24}$$

Again, from Eqs. (5.17) and (5.23), we have

$$y^2 = s^2 + c^2. \tag{5.25}$$

The relations between four variables, x, y, s and ψ, involved in the theory of strings, have been deduced above and are reproduced in Table 5.1 in compact form for students' convenience. They are advised to commit to memory various relations given in this table as a judicial application of these will facilitate the solution of the common catenary problems. The remaining variable involved is the tension T and the student shall have no difficulty in arriving at following relations:

$$\text{Tension: } T = w\,y \tag{5.26}$$

$$\text{Horizontal component: } T_0 = w\,c, \tag{5.27}$$

$$\text{Vertical component: } T_v = w\,s. \tag{5.28}$$

Table 5.1 Relations between four variables x, y, s, and ψ.

x	x	$c \cosh^{-1}\left(\dfrac{y}{c}\right)$	$c \sinh^{-1}\left(\dfrac{s}{c}\right)$	$c \log(\sec\psi + \tan\psi)$
y	$c \cosh\left(\dfrac{x}{c}\right)$	y	$\sqrt{s^2 + c^2}$	$c \sec\psi$
s	$c \sinh\left(\dfrac{x}{c}\right)$	$\sqrt{y^2 - c^2}$	s	$c \tan\psi$
ψ	$\tan^{-1} \sinh\left(\dfrac{x}{c}\right)$	$\sec^{-1}\left(\dfrac{y}{c}\right)$	$\tan^{-1}\left(\dfrac{s}{c}\right)$	ψ

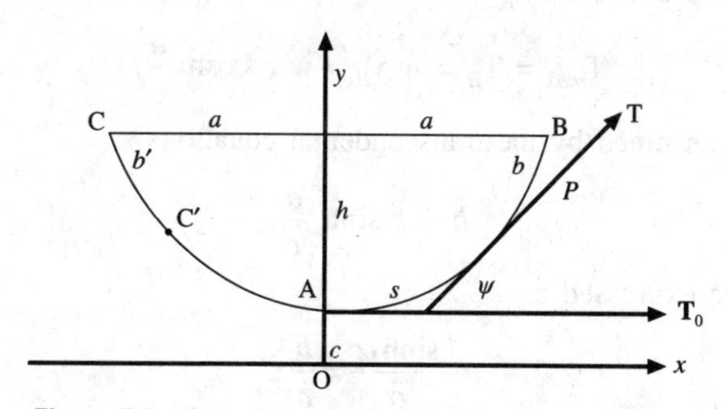

Figure 5.3 Span $2a$ and sag h for a common catenary.

The two points B and C, from which the chain or the string is hung, determine the span $2a$ of the catenary if they are at the same level. The catenary then is symmetrical about the y-axis which is the axis of the catenary. When the two supporting points B and C′ are not at the same level, the curve can be extended by adding the portion C′C (Figure 5.3) and still considered as a part of the catenary with span BC. The depth h of the vertex of the catenary below the span BC is known as sag. Designating by subscript the point at which the quantity is being considered, we can write

$$\text{Span: } 2a = 2x_B, \tag{5.29}$$

$$\text{Sag: } h = y_B - y_A = y_B - c, \tag{5.30}$$

$$\text{Length: } 2b = 2s_B. \tag{5.31}$$

Applications of Eqs. (5.29 and 5.31) to the solution of two important problems concerning freely hanging strings are as follows:

Application (i) To find the span when the length and sag are given: This problem is of frequent occurrence, e.g., when the distance between two points B and C is being measured by a measuring chain or tape which sags under its own weight. We obtain

$$s_B = b,\ y_B = h + c,\ x_B = a. \tag{5.32}$$

Now, using the formulae $\quad y_B{}^2 = s_B{}^2 + c^2$

we get
$$c = \frac{b^2 - h^2}{2h}. \tag{5.33}$$

The span is given by substituting above values in Eq. (5.24). Thus,

$$\text{Span} = 2a = 2x_B = 2c \log \frac{y_b + s_B}{c} = \frac{b^2 - h^2}{h} \log \frac{b+h}{b-h}. \tag{5.34}$$

Application (ii) To find the maximum tension of a cable with given span and length: From Eq. (5.26), it is clear that the maximum tension will be at the highest point B, and there $x_B = a$; hence

$$T_{max} = T_B = w\ y_B = w\ c \cosh \frac{a}{c}, \tag{5.35}$$

where c is determined by the transcendental equation

$$b = c \sinh \frac{a}{c},$$

which may be expressed as

$$\frac{\sinh \alpha}{\alpha} = \frac{b}{a} \tag{5.36}$$

with c given by $\frac{a}{\alpha}$. Knowing b and a, the value of α may be obtained from the numerical tables of $\frac{\sinh \alpha}{\alpha}$.

5.4 APPROXIMATIONS TO COMMON CATENARY

Many-a-times in applied problems of practical interest there is no material difference if we use approximate results, except for the advantage that then the problem becomes easier to handle. Following are the two useful approximations to common catenary:

(i) *Exponential approximation:* The catenary is

$$y = c \cosh \frac{x}{c} = \frac{1}{2} c\,(e^{\frac{x}{c}} + e^{\frac{-x}{c}}).$$

When $\frac{x}{c}$ is large, i.e., for points far away from the vertex $e^{\frac{-x}{c}} << e^{\frac{x}{c}}$, and then we have the exponential approximation as

$$y \approx \frac{1}{2} c e^{\frac{x}{c}}. \tag{5.37}$$

(ii) *Parabolic approximation:* When c is large or for points near the vertex, $\frac{x}{c}$ is small; and then we have the approximation as

$$y \approx c + \frac{x^2}{2c}, \tag{5.38}$$

on neglecting $O\left(\frac{x^2}{c^2}\right)$ terms in the expansion of $\cosh \frac{x}{c}$. Equation (5.38) represents a parabola of latus rectum $2c$.

It may be noticed that since radius of curvature

$$\rho = \frac{\mathrm{d}s}{\mathrm{d}\psi} = c \sec^2 \psi,$$

c will be large in the case of a tightly stretched string as ρ is large. This may also be inferred from Eq. (5.36) which is approximated by

$$\frac{\sinh \alpha}{\alpha} \approx 1 \tag{5.39}$$

as in the case of tightly stretched string $\frac{b}{a} \approx 1$. Hence, $\alpha = 0$ and so $c = \frac{a}{\alpha}$ is large in the case of a tightly stretched string.

Using the relation $s = c \sinh\left(\frac{x}{c}\right)$ we can also write the approximation as

$$s \approx x + \frac{x^3}{6c^2} \tag{5.40}$$

giving the increase in the length of the string due to sag as

$$2(b-a) \approx \frac{a^3}{3c^2}. \tag{5.41}$$

Also, we have from Eq. (5.38)

$$h = \text{Sag in the middle} = (y-c)|_{x=a}$$

$$\approx \frac{a^2}{2c} \approx \frac{b^2}{2c}. \tag{5.42}$$

It generally happens that the measuring chain or the cable in Applications (i) and (ii) section (5.3.1) are held in a tightly stretched state and then, using Eq. (5.42) in Eqs. (5.34) and (5.35), it is not difficult to show that

$$\text{Span} = 2a \approx 2b\left(b - \frac{2h^2}{3b^2}\right), \tag{5.43}$$

$$T_{max} = wc \approx wa\sqrt{\frac{a}{6(b-a)}}. \tag{5.44}$$

5.5 CATENARY OF UNIFORM STRENGTH

The curve in which a string hangs when its line mass density m at each of its points is proportional to the tension there is known as catenary of uniform strength. Thus, in the case the curve is a plane curve (?) and we have

$$f_x = 0, f_y = -g, m = \lambda T, \tag{5.45}$$

where λ is some constant. Equation (5.8), then, provides

$$T\frac{\mathrm{d}x}{\mathrm{d}s} = \text{Horizontal component of tension} = \text{Constant} = T_0 \tag{5.46}$$

and

$$\frac{\mathrm{d}}{\mathrm{d}s}\left(T\frac{\mathrm{d}y}{\mathrm{d}s}\right) = \lambda g T. \tag{5.47}$$

Equations (5.46) and (5.47) can be manipulated to yield the following differential equation for catenary of uniform strength:

$$a\frac{\mathrm{d}^2 y}{\mathrm{d}x^2} = 1 + \left(\frac{\mathrm{d}y}{\mathrm{d}s}\right)^2 \tag{5.48}$$

where $a = \dfrac{1}{\lambda g}$.

Writing $p = \dfrac{\mathrm{d}y}{\mathrm{d}x}$, Eq. (5.48) becomes

$$\frac{dp}{1+p^2} = \frac{dx}{a},$$

on integration which provides

$$\frac{dy}{dx} = p = \tan\left(\frac{x}{a} + A\right). \tag{5.49}$$

If we choose the lowest point of the curve as lying on the y-axis i.e., $\frac{dy}{dx} = 0$ when $x = 0$, we get A = 0, and consequently

$$\frac{dy}{dx} = p = \tan\frac{x}{a}.$$

Integrating Eq. (5.49), we obtain

$$y = a \log \sec\frac{x}{a}, \tag{5.50}$$

the constant of integration vanishing if we choose $y = 0$ when $x = 0$, i.e. if we assume the string to pass through origin. Equation (5.50) is the Cartesian equation of the catenary of uniform strength. Noticing that $\frac{dy}{dx} = \tan\psi$, Eq. (5.49) immediately gives

$$x = a\,\psi. \tag{5.51}$$

Now $\frac{ds}{d\psi} = \frac{ds}{dx}\cdot\frac{dx}{d\psi} = a\sec\psi$, and so on integration we get the intrinsic equation

$$s = a\log(\sec\psi + \tan\psi), \tag{5.52}$$

where we have taken $s = 0$ when $\psi = 0$. Equation (5.52) may also be expressed as

$$\cosh\frac{s}{a} = \sec\psi, \tag{5.53}$$

Further, the radius of curvature is

$$\rho = \frac{ds}{d\psi} = a\sec\psi, \tag{5.54}$$

and this together with Eq. (5.46) easily leads to the result

$$m = \lambda T = \lambda T_0 \sec\psi = \lambda T_0 \frac{\rho}{a},$$

i.e., the mass per unit length at any point in a catenary of uniform strength varies as the corresponding radius of curvature. The converse of this viz., if the mass density of string varies as the radius of curvature of the curve in which it hangs then this curve is a catenary of uniform strength, is also true (the student is advised to prove it).

5.6 SUSPENSION BRIDGE

Let us consider the cable AB of a suspension bridge which is under a continuous force due to a uniformly distributed load on a horizontal line A′B′ representing the roadway of weight w_0 per unit length (Figure 5.4). The weight of the cable and the vertical rods supporting the roadway is assumed to be negligible as compared to the weight of the roadway.

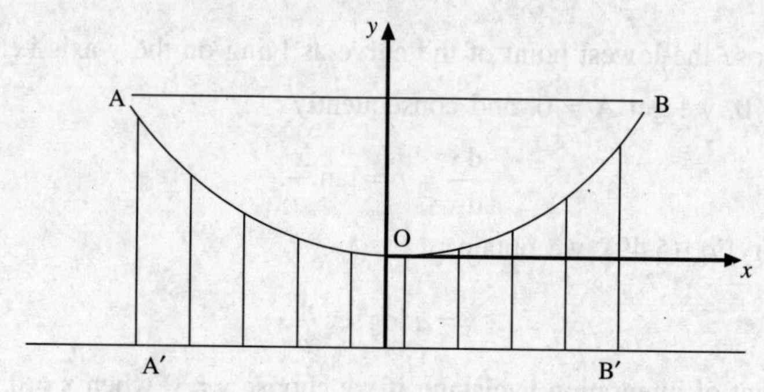

Figure 5.4 Road *A′B′* suspended by cable AB.

Thus, the force on length d*s* of the cable AB is w_0 d*x* acting vertically downwards. Hence, from Eq. (5.8), we take

$$mf_x = 0, mf_y = -w_0 \frac{dx}{ds}, \tag{5.55}$$

to obtain

$$\frac{d}{ds}\left(T \frac{dx}{ds}\right) = 0, \tag{5.56}$$

$$\frac{d}{ds}\left(T \frac{dy}{ds}\right) = w_0 \frac{dx}{ds}. \tag{5.57}$$

From Eq. (5.56), we get

$$T \frac{dx}{ds} = \text{constant} = T_0. \tag{5.58}$$

which on being substituted in Eq. (5.57) provides the differential equation as

$$\frac{d^2 y}{dx^2} = \frac{w_0}{T_0}. \tag{5.59}$$

This may be recognized as the differential equation of a parabola yielding the solution

$$y = \frac{w_0}{2T_0} x^2 \tag{5.60}$$

when origin is chosen as the lowest point of the cable, i.e., if we take $y = 0$, $\frac{dy}{dx} = 0$ when $x = 0$.

The tension in the cable as given by Eq. (5.58) is

$$T = T_0 \frac{ds}{dx} = \sqrt{T_0^2 + w_0^2}. \tag{5.61}$$

5.7 STRINGS IN CONTACT WITH CURVES

So far we have studied about unconstrained strings. Let us consider the case of a heavy string lying in contact with a surface along a plane curve. The surface in general will be considered to be rough but, taking the coefficient of friction $\mu = 0$, the results for a smooth contact may be derived. We shall assume the external force system to be the force due to gravity (represented by w, the weight per unit length of the string) and incorporate the forces of constraint, viz., the normal reaction R and the force of friction μR, also as external forces.

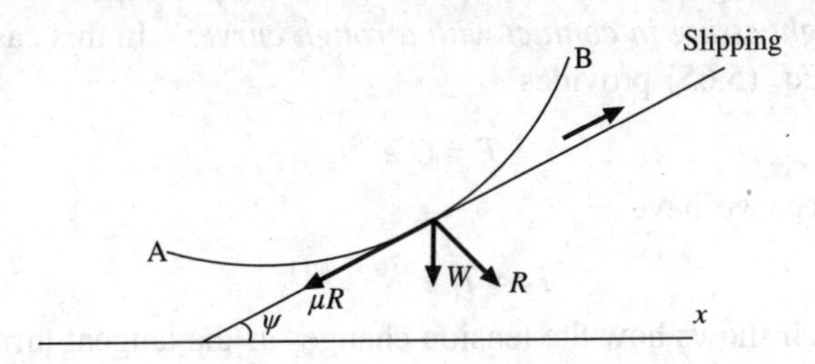

Figure 5.5 Heavy string AB in contact with a rough surface.

Assuming the slipping to have a tendency in the direction of s increasing (Figure 5.5), the differential equations of equilibrium Eqs. (5.6) and (5.7) may be expressed as

$$\text{Tangential: } \frac{1}{\rho}\frac{dT}{d\psi} - w\sin\psi - \mu R = 0, \tag{5.62}$$

$$\text{Normal: } \frac{T}{\rho} - w\cos\psi - R = 0, \tag{5.63}$$

where ψ as usual is the slope of tangent with the horizontal x-axis and $\rho = \frac{ds}{d\psi}$, the radius of curvature.

Eliminating R in between Eqs. (5.62) and (5.63), we get

$$\frac{dT}{d\psi} - \mu T = w\rho\,(\sin\psi - \mu\cos\psi). \tag{5.64}$$

The solution of the above linear differential equation of the first order can easily be written as

$$Te^{-\mu\psi} = C + \int w\rho\,(\sin\psi - \mu\cos\psi)\, e^{-\mu\psi}\, \mathrm{d}\psi. \qquad (5.65)$$

The solution may be completed by evaluating the constant of integration C from the end condition pertaining to the problem. Knowing T, we obtain the value of R from Eq. (5.63).

Giving suitable values to m and w, we discuss some special cases as follows:

(i) *A light string in contact with a smooth curve:* In this case, $\mu = w = 0$, hence Eq. (5.65) shows that

$$T = \text{constant} = T_0, \qquad (5.66)$$

and then we have the reaction as

$$R = \frac{T_0}{\rho}. \qquad (5.67)$$

Equation (5.66) embodies the frequently used result of constancy of tension in a light string. Equation (5.67) explains the greater tendency of a string to bite into a parcel at a sharper edge.

(ii) *A light string in contact with a rough curve:* In this case $w = 0$, and so Eq. (5.65) provides

$$T = C\, e^{\mu\psi}, \qquad (5.68)$$

hence, we have

$$T_2 = T_1\, e^{\mu(\psi_2 - \psi_1)}, \qquad (5.69)$$

which shows how the tension changes as the tangent turns through an angle $\psi_2 - \psi_1$. It is to be noted that in Eq. (5.69) $\psi_2 - \psi_1$ is positive for change in the direction of slipping.

The exponential increase of tension with increasing ψ is of great practical importance, e.g., in holding ships by ropes passing round moving posts and in hoists. Student can show that a weight of 2 kg tied to one end of a rope going twice round a post, having coefficient of friction $\frac{1}{2}$, can sustain a load of more than 1 ton.

(iii) *A heavy string in contact with a smooth curve:* Now we have $\mu = 0$, Therefore, Eq. (5.65) becomes

$$T = C + \int w\, \mathrm{d}y, \qquad (5.70)$$

where we have made use of the result $\rho = \dfrac{\mathrm{d}s}{\mathrm{d}\psi}$, $\sin\psi = \dfrac{\mathrm{d}y}{\mathrm{d}s}$.

If the chain or the uniform string is in contact with a smooth curve, the difference of the tensions at any two points is equal to the weight of a portion of the string whose length is equal to the vertical distance between the two points.

SOLVED EXAMPLES

EXAMPLE 5.1 A uniform string AB of length b and weight wb, will break if subjected to a tension in excess of $2wb$. It is fixed at B and is in equilibrium under the action of horizontal force F at A. Show, if the string is just about to break, that $F = wb\sqrt{3}$ and find the horizontal and vertical distances of B from A.

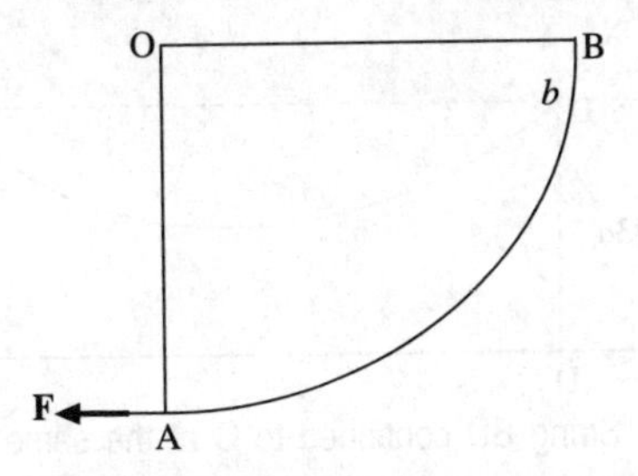

Figure 5.6 String AB with horizontal force F at A.

Solution Since tension at A is equal to F, it is horizontal there, and so A is the vertex of the catenary in which the string AB hangs. Thus, we have

$$F = T_0 = wc. \tag{i}$$

Further, as the maximum tension occurs at the highest point B, we have

$$wy_B = T_B = 2wb,$$

so that

$$y_B = 2b.$$

Now, using the relation $y_B{}^2 = s_B{}^2 + c^2$, with $s_B = b$, we obtain $c = b\sqrt{3}$ and this substituted in Eq. (i) provides the required value $F = wb\sqrt{3}$.

The vertical distance is obviously $2b$ and the horizontal distance easily follows on using Eq. (5.24).

EXAMPLE 5.2 A uniform chain of length $2b$, which can bear a tension n times its weight, is stretched between two points in the same horizontal line. Show that the least possible sag in the middle is $2b\left[n - \sqrt{\left(n^2 - \frac{1}{4}\right)}\right]$.

Solution We have $s_B = b$, and so

$$h = \text{Sag in the middle} = y_B - c$$

$$= y_B - \sqrt{(y_B^2 - b^2)}. \tag{i}$$

From Eq. (i) we find that $\frac{dh}{dy_B} < 0$. Thus, h is a decreasing function of y_B, and hence, it is minimum for maximum value of y_B. Since the tension is maximum at the highest point B and is given not to exceed $2wnb$, we conclude that for minimum sag $y_B = 2nb$. Substituting this in (1), we get the required answer.

EXAMPLE 5.3 A uniform flexible chain, of length $54a$, hangs over two small smooth pegs, the lengths of the vertical portions being $20a$ and $13a$, respectively. Show that the parameter of the catenary between the pegs is $12a$ and that the horizontal distance between the pegs is $12a \log \left(\frac{9}{2}\right)$.

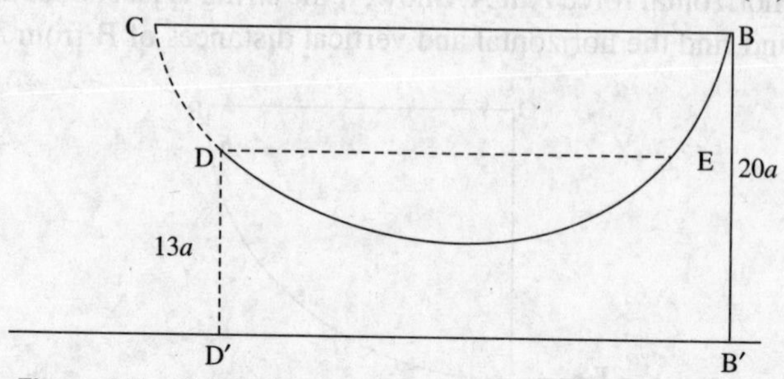

Figure 5.7 String BD continued to C at the same level as B.

Solution The continuity of the tension at B and D (as shown in Figure 5.7) provides

$$y_B = \frac{T_B}{w} = BB' = 20a, \; y_D = \frac{T_D}{w} = DD' = 13a. \tag{i}$$

Since y_B is the height above the directrix, this also show that the ends B′ and D′ of the vertical positions lie on the directrix B′D′.

Suppose $s_B = b$, as measured from the vertex A, then

$$s_D = s_E = 54a - 20a - 13a - b = 21a - b. \tag{ii}$$

Using the values given in Eq. (ii) in the relation $y^2 = s^2 + c^2$,we obtain

$$400a^2 = b^2 + c^2$$

$$169a^2 = (21a - b)^2 + c^2.$$

The above two equations provide $c = 12a$ and then $s_B = b = 16a$ and $s_D = 21a - b = 5a$. These values give the horizontal distance between the pegs

$$= x_B + |x_D| = c\left[\log \frac{y_B + s_B}{c} + \log \frac{y_D + s_D}{c}\right]$$

$$= c \log \frac{20+16}{13+5} = 12a \log \frac{9}{2}.$$

EXAMPLE 5.4 Show that for an in extensible string lying in equilibrium in a plane under a central force **F** per unit mass

$$T = \int m\mathbf{F}\, \mathrm{d}r + \text{Constant}, \; T\,p = \text{Constant}$$

where T is the tension, p the perpendicular from origin on the tangent and m the mass per unit length per unit mass.

Solution Let the inextensible string be subjected to a central force **F** per unit mass. **F** is a function of the radial distance r from the pole O which is taken as the centre of repulsion (refer Figure 5.8).

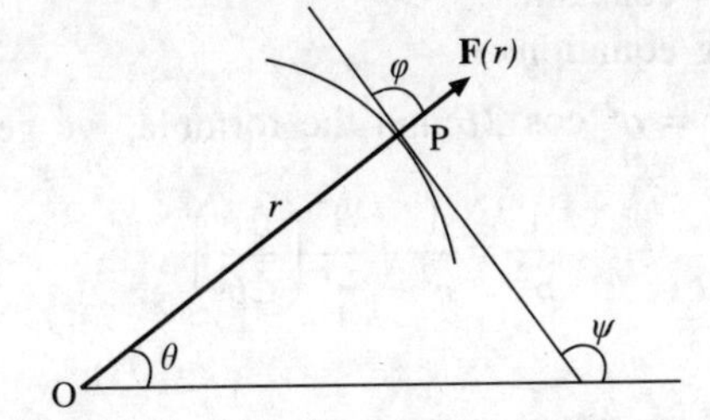

Figure 5.8 String AB under central force F(r).

Thus, the tangential and normal components of the force **F** at P(r, θ) are $F\cos\varphi$ and $-F\sin\varphi$, respectively, where $\tan\varphi = r\dfrac{d\theta}{dr}$, and the differential Eqs. (5.6) and (5.7) assume the form

$$\frac{dT}{ds} + mF\cos\varphi = 0, \tag{i}$$

$$\frac{T}{\rho} - mF\sin\varphi = 0. \tag{ii}$$

Since $\cos\varphi = \dfrac{dr}{ds}$, Eq. (i) provides

$$dT = -\,mF\,dr, \tag{iii}$$

giving

$$T = -\int mF\,dr + \text{Constant}. \tag{iv}$$

as required.

Further, we have in pedal form

$$\rho = r\frac{dr}{dp},\; p = r\sin\varphi, \tag{v}$$

which, on using Eq. (iii) when introduced in Eq.(ii) yields

$$Tdp = -\,pdT$$

giving

$$Tp = \text{Constant}. \tag{vi}$$

EXAMPLE 5.5 Find the law of force in the case of a string resting in form of following curve under a central force F

(i) Leminscate, $r^2 = a^2\cos 2\theta$
(ii) Equiangular spiral, $r = a\,e^{\theta\cot\alpha}$

Solution

(i) We know from Example 5.4 that

$$F = \frac{K}{mp^2}\frac{dp}{dr}, \tag{i}$$

where, K = constant.

Now, using equation

$r^2 = a^2 \cos 2\theta$ and the formula, we get

$$\frac{1}{p^2} = \frac{1}{r^2} + \frac{1}{r^4}\left(\frac{\mathrm{d}r}{\mathrm{d}\theta}\right)^2,$$

we get

$$\frac{1}{p} = \frac{a^2}{r^3},$$

which when substituted in the formula given in Eq. (i) gives

$$F = \frac{3\mathrm{K}a^2}{mr^4}.$$

(ii) Similarly show that in this case $F \propto \dfrac{1}{r^2}$.

EXAMPLE 5.6 Four rough circular pegs are at the angular points of a square in a vertical plane with its sides horizontal and vertical. Over each peg passes a string supporting a weight W, and the other ends of these four strings are knotted together. Show that the greatest weight that can be attached to this knot so that it may remain in equilibrium at the centre of the square is $2\sqrt{2}\, W\, e^{\frac{\mu\pi}{4}} \sinh\left(\dfrac{\mu\pi}{2}\right)$.

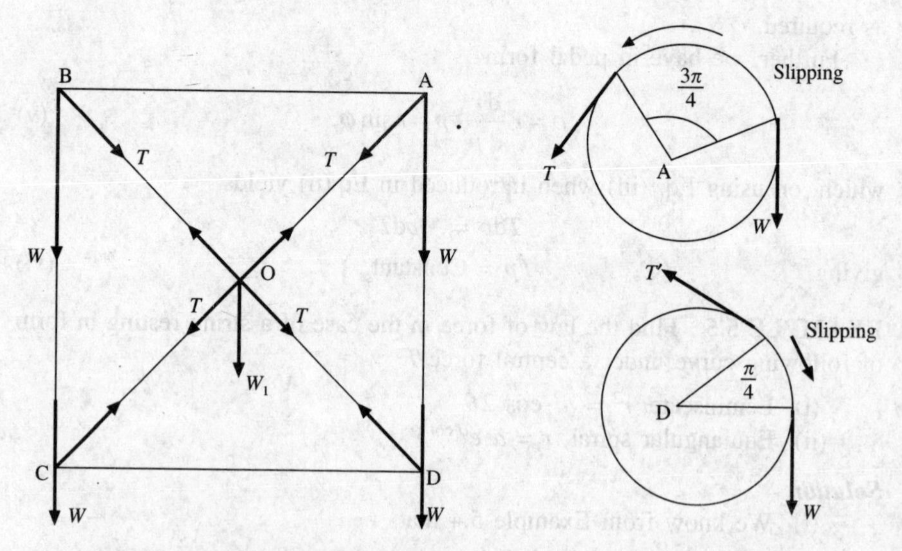

Figure 5.9 Rough pegs at the four corners A, B, C, D of a square and the strings knotted at its centre O.

Solution Let A, B, C, D be the four pegs and W_1 the weight tied to the knot at the centre O. W_1 will be maximum when it has a tendency to come down, and then the direction of the slipping are as marked in the blown up diagrams of an upper peg A and a lower peg D. The system is symmetrical about vertical through O and the forces are as marked in Figure 5.9. Notice that at the peg A, the tangent has negotiated an angle $\frac{3\pi}{4}$ and at the peg D an angle $\frac{\pi}{4}$. Keeping in mind the relation between the change in direction of the tangent and the direction of slipping and using Eq. 5.69, we have

$$T = We^{\frac{3\mu\pi}{4}}, \; T' = We^{\frac{-\mu\pi}{4}}.$$

Also, the equilibrium condition at O gives

$$W = 2(T - T')\cos\frac{\pi}{4}.$$

Substituting the values of T and T' and simplifying we get the required result.

EXAMPLE 5.7 A heavy uniform string is placed on a rough catenary whose axis is vertical and vertex upwards; the coefficient friction being given by $e^{\mu\pi} = 4$. Show that the string will lie in limiting equilibrium with one end at the vertex if its length is equal to the parameter of the catenary.

Solution In the position of limiting equilibrium the string will be on the point of slipping down and the maximum frictional force μR will be acting as marked in Figure 5.10, R being the normal reaction.

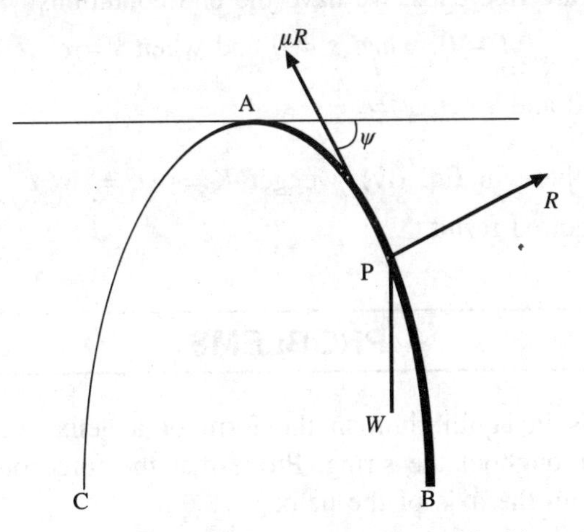

Figure 5.10 String AB lying on a rough catenary BC with vertex at A.

The external force is the weight w. All these forces are per unit length acting at a point P of the string A of length c with the end A at the vertex of the catenary. Then, we have

$$mf_t = w \sin \psi - \mu R,$$

$$mf_n = w \cos \psi - R.$$

Substituting these values in Eqs. (5.6) and (5.7), we get following differential equations of equilibrium

Tangential: $$\frac{1}{\rho}\frac{\mathrm{d}T}{\mathrm{d}\psi} - \mu R = -w \sin\psi, \qquad \text{(i)}$$

Normal: $$\frac{T}{\rho} - R = -w \cos\psi. \qquad \text{(ii)}$$

The given curve being the catenary $s = c \tan \psi$ and so we have $\rho = \frac{\mathrm{d}s}{\mathrm{d}\psi}$ $= c \sec^2 \psi$. Substituting this value of ρ in Eq. (ii) and eliminating R in between Eq. (i) and Eq. (ii), we have

$$\frac{\mathrm{d}T}{\mathrm{d}\psi} - \mu T = wc \sec^2 \psi \, (\mu \cos\psi - \sin\psi). \qquad \text{(iii)}$$

Integrating the linear differential equation (iii) with the

$$Te^{-\mu\psi} = K + wc \int e^{-\mu\psi} \, (\mu \sec\psi - \sec\psi \tan\psi) \, \mathrm{d}\psi$$

$$= K - wc\, e^{-\mu\psi} \sec\psi. \qquad \text{(iv)}$$

Since A and B are free ends, we have the end conditions

$$T = 0, \text{ when } s = 0 \text{ and when } s = c \qquad \text{(v)}$$

i.e., when $\psi = 0$ and when $\psi = \frac{\pi}{4}$.

Applying these in Eq. (iv), we get $K = wc = wc\, e^{-\frac{\mu\pi}{4}} \sec\frac{\pi}{4}$, which provides the required result.

PROBLEMS

1. A string is in equilibrium in the form of a helix, and the tension is constant throughout the string. Prove that the force on any element is directly from the axis of the helix.
2. In a non-uniform string hanging under gravity the area of the cross section at any point is inversely proportional to tension. Show that the curve is an arc of a parabola with axis vertical.
3. A heavy string is suspended from two points, and the density is such that the form of the string is an equiangular spiral. Show that the density at

any point P is inversely proportional to $r \cos^2 \psi$, where r is the distance of P from the pole and ψ the angle the tangent at P makes with the horizontal.

4. If a uniform string hangs in the form of a parabola whose focus is S, under the action of normal forces only show that the force at any point P varies inversely as $(SP)^{\frac{3}{2}}$.

5. A heavy uniform chain AB hangs freely, under gravity, with the end A fixed and the other end B attached by a light string BC to a fixed point C at the same level as A. The lengths of the string and chain are such that the ends of the chain at A and B make angles of 60° and 30°, respectively with the horizontal. Prove that the ratio of these lengths is $(\sqrt{3}-1):1$.

6. A heavy chain, of length $2l$, has one end tied at A and the other is attached to a small heavy ringwhich can slide on a rough horizontal rod which passes through A. If the weight of the ring be n times the weight of the chain, show that its greatest possible distance from A is

$$\frac{2l}{\lambda} \log\left[\lambda + \sqrt{1+\lambda^2}\right],$$

where $\dfrac{1}{\lambda} = \mu(2n+1)$ and μ is the coefficient of friction.

7. A uniform chain, of length l and weight W, hangs between two fixed points at the same level, and weight W' is attached at the middle point. If k be the sag in the middle, prove that the pull on either point of support is

$$\frac{k}{2l}W + \frac{l}{4k}W' + \frac{l}{8k}W.$$

8. A uniform chain AB of length $l = a + b$ hangs from the end B with a portion AP of length a resting on a smooth plane inclined at an angle α to the horizontal. Prove that the height of B above the level of A is h, where $h^2 = l^2 \sin^2 \alpha + b^2 \cos^2 \alpha$.

9. A box kite is flying at a height h with a length l of wire paid out, and with the vertex of the catenary on the ground. Show that at the kite the inclination of the wire to the ground is $2\tan^{-1}\left(\dfrac{h}{l}\right)$, and that its tensions there and at the ground are

$$w\frac{l^2+h^2}{2h} \text{ and } w\frac{l^2-h^2}{2h},$$

where w is the weight of the wire per unit of length.

10. The sag of a telegraph wire stretched between two poles at a distance $2a$ apart is k. Show that the tension at each end of the wire is approximately

$$w\left(\frac{a^2}{2k}+\frac{7k}{6}\right),$$

where w is the weight per unit length of the wire.

11. A heavy string hangs over two fixed small smooth pegs. The two ends of the strings are free and the central portion hangs in a catenary. Show that the free ends of the string are on the directrix of the catenary. If the two pegs are on the same level and distance $2a$ apart, show that equilibrium is impossible unless the length of the string is equal to or greater than $2ae$.

12. An endless uniform chain is hung over two smooth pegs in the same horizontal line. Show that, when it is in a position of equilibrium, the ratio of the distance between the vertices of the two catenaries to half the length of the chain is the tangent of half the angle of inclination of the portions near the pegs.

13. A uniform chain of length l hangs between two points A and B which are at a horizontal distance a from one another, with B at a vertical distance b above A. Prove that the parameter of the catenary is given by

$$2c\sinh\left(\frac{a}{2c}\right)=\sqrt{(l^2-b^2)}.$$

Prove also that, if the tensions at A and B are T_1 and T_2 respectively

$$T_1+T_2=W\sqrt{\left(1+\frac{4c^2}{l^2-b^2}\right)}$$

and
$$T_2-T_1=\frac{Wb}{l},$$

where W is the weight of the chain.

14. A length l of a uniform chain has one end fixed at a height h above a rough table, and rests in a vertical plane so that a portion of it lies in a straight line on the table. Prove that if the chain is on the point of slipping, the length on the table is

$$l+\mu h-\sqrt{\{(\mu^2+1)h^2+2\mu lh\}},$$

where μ is the coefficient of friction.

15. The end links of a uniform chain slide along a fixed rough horizontal rod. Prove that the ratio of the maximum span to the length of the chain is

$$\mu\log\frac{1+\sqrt{(1+\mu^2)}}{\mu},$$

where μ is the coefficient of friction.

16. A heavy chain, of total length l, rests with one end on a rough table and the other on a rough floor. It is stretched out as far as possible so that the equilibrium is limiting; if μ be the coefficient of friction both at the table and the floor and b the height of the table, show that the length on the floor is

$$\frac{\mu}{4}\left\{b+\frac{2l}{\mu}-\sqrt{\left(b^2+\frac{4lb}{\mu}\right)}\right\}-\frac{b}{2\mu},$$

provided that $l > b + \left(\frac{b}{\mu}\right)$.

17. One extremity of a uniform string is attached to a fixed point, and the string rests partly on a smooth inclined plane; prove that the directrix of the catenary determined by the portion which is not in contact with the plane is the horizontal line drawn through the extremity which rests on the plane.

18. A uniform string of weight W is suspended from two points at the same level and a weight W_1 is attached to its lowest point. If α and β are now the inclinations to the horizontal of the tangents at the highest and lowest points, prove that

$$\frac{\tan\alpha}{\tan\beta}=1+\frac{W}{W_1}.$$

19. A heavy uniform string, 90 cm long, hangs over two smooth pegs at different heights. The parts which hang vertically are of lengths 30 and 33 cm. Prove that the vertex of the catenary divides the whole string in the ratio 4:5, and find the distance between the pegs.

[**Ans:** Horizontal distance between the pegs

$$=20\sqrt{2}\left[\cosh^{-1}\left(\frac{3}{4}\sqrt{2}\right)+\cosh^{-1}\left(\frac{33}{40}\sqrt{2}\right)\right]]$$

20. Show that the length of an endless chain which will hang over a circular pulley of radius a so as to be in contact with two-thirds of the circumference of the pulley is

$$a\left[\frac{3}{\log\left(2+\sqrt{3}\right)}+\frac{4\pi}{3}\right].$$

21. A uniform heavy chain of length $2l$ is stretched between two supports at the same horizontal level. The lowest point of the chain is at a depth h below the supports. Prove that the centre of gravity of the chain is at a depth

$$\frac{h^2+l^2}{4h}-\frac{(l^2-h^2)^2}{8h^2l}\sinh^{-1}\frac{2hl}{l^2-h^2}$$

22. A telegraph wire is made up of a given material and such a length l is stretched between two posts, distant d apart and of the same height, as will produce the least possible tension at the posts. Show that $l = \left(\frac{d}{\lambda}\right)\sinh\lambda$, where λ is given by the equation $\lambda\tanh\lambda = 1$.

23. The end links of a uniform chain of length l can slide on two smooth rods in the same vertical plane which are inclined in opposite directions at equal angles θ to the vertical. Prove that the sag in the middle is $\left(\frac{l}{2}\right)\tan\frac{\theta}{2}$.

24. The end links of a uniform chain of length $2l$ can slide on two smooth rods in the same vertical plane which are inclined in opposite directions at equal angles α to the vertical. Prove that the distance between the links is $2l\cot\alpha\sinh^{-1}(\tan\alpha)$.

25. A string rests in the form of a plane curve under the action of central repulsive force; if the force at any point be proportional to the curvature, prove that the curve is a parabola.

26. If a string is in equilibrium under any central force, then prove that the resultant actions of these central forces on any portion PQ of the string is along OA, where O is the centre of force and A the point of intersection of the tangents at P and Q.

27. Three equally rough pegs A, B, C of the same circular cross section are placed at the corners of an equilateral triangle, so that BC is horizontal and A above BC. Show that the greatest weight which can be supported by a weight W tied to the end of a string, which is carried once round the pegs and does not completely surround each peg, is $We^{5\mu\pi}$, where μ is the coefficient of friction.

28. A light string is passed over two rough pegs A and B in the same horizontal line at a distance $2a$ apart. The ends are fastened to a weight C, and in the position of limiting equilibrium AB subtends a right angle at C. Show that the horizontal distance of C in this position from the middle of AB is $a\tanh\left(\frac{3\mu\pi}{2}\right)$, where μ is the coefficient of friction.

29. A uniform heavy string rests on a smooth parabola, whose axis is vertical and vertex upwards, so that its ends are extremities of the latus rectum. Show that the pressure on the curve, at the point where the tangent makes an angle ϕ with horizontal, is

$$\frac{w}{2}(2\cos^3\phi + \cos\phi),$$

where w is the weight of the string per unit length.

30. A heavy uniform string rests symmetrically in a smooth catenary whose axis is vertical and vertex upwards; find tension and pressure on the curve at any point.

31. A string is placed upon a rough circle which is fixed vertically. The string subtends an angle β at the centre. If the string is on the point of slipping off, prove that the angular distance α of its upper end from the highest point of the circle is determined by the equation.

$$\frac{\cos(\alpha + \beta - 2\varepsilon)}{\cos(\alpha - 2\varepsilon)} = \beta \tan \varepsilon,$$

where ε is the angle of friction and α is measured in the direction towards which the string is slipping.

32. A uniform heavy string rests on the upper surface of a vertical circle of radius a, and partly hangs vertically. Show that, if one end be at the highest point of the circle, the greatest length that can hang freely is

$$\frac{2\mu a + (\mu^2 - 1)\, a e^{\frac{\mu\pi}{2}}}{\mu^2 + 1}.$$

33. A string rests on a rough semicircle, being acted on constant attractive force towards one of its extremities, the friction is just sufficient to prevent motion. Show the coefficient of friction is given by

$$e^{\mu\pi} = \frac{3\mu}{1 - 2\mu^2}.$$

34. A heavy uniform chain rests on a rough cycloid whose axis is vertical and vertex upwards, one end of the chain being at the vertex and the other at a cusp; if the equilibrium be limiting, show that

$$(1 + \mu^2)\, e^{\frac{\mu\pi}{2}} = 3.$$

35. A uniform inextensible string, of length l, hangs in equilibrium over a fixed rough cylinder of radius a, whose axis is horizontal; show that the length of the greater of the vertical portion is

$$\frac{l - \pi a}{1 + e^{-\mu\pi}} + \frac{2\mu a}{1 + \mu^2}.$$

6 Rectilinear Motion

6.1 INTRODUCTION

Although motion in a straight line or rectilinear motion constitute the simplest of dynamical problems, yet it is very important because many physical problems reduce to this category, e.g., simple harmonic motion, motion under inverse square law, motion in a resisting medium and motion of a rocket. Therefore, in this chapter, we first proceed to determine the solution of the one-dimensional equation of motion as given by Newton's second law and subject to initial conditions.

6.2. EQUATION OF MOTION AND ITS INTEGRATION

For a particle moving in a straight line (say x-axis), there is only one space variable x, only one component of velocity given by

$$v = \frac{\mathrm{d}x}{\mathrm{d}t}, \tag{6.1}$$

and only one component of acceleration given by

$$f = \frac{\mathrm{d}^2 x}{\mathrm{d}t^2} = \frac{\mathrm{d}v}{\mathrm{d}t} = v\frac{\mathrm{d}v}{\mathrm{d}x} = \frac{1}{2}\frac{\mathrm{d}v^2}{\mathrm{d}x}. \tag{6.2}$$

The total external force $\mathrm{F}(x, v, t)$ in general, is a function of position x, velocity v and time t, and from Newton's second law we have a single equation of motion, given as

$$m\frac{\mathrm{d}v}{\mathrm{d}t} = m\frac{\mathrm{d}^2 x}{\mathrm{d}t^2} = \mathrm{F}(x, v, t), \tag{6.3}$$

m being the mass of the particle.

Many a times, the particle is taken to be of unit mass or m may be absorbed in F(F = mf), and then Eq. (6.3) becomes

$$\frac{\mathrm{d}v}{\mathrm{d}t}=\frac{\mathrm{d}^2 x}{\mathrm{d}t^2}=f(x, v, t); \tag{6.4}$$

here a may be recognized as acceleration.

Equations (6.3) or (6.4) are second order differential equations, and for complete specification of the problem two initial conditions are needed which generally are prescribed as

$$x = x_0,\ \frac{\mathrm{d}x}{\mathrm{d}t} = v = v_0 \text{ at } t = 0. \tag{6.5}$$

The two arbitrary constants A and B in the solution

$$x = x\ (t,\ \mathrm{A},\ \mathrm{B}) \tag{6.6}$$

of Eq.(6.4) are, thus, obtained from the two conditions given in Eq. (6.5).

The simple looking equation of motion is deceptive in its appearance as its solution may entail, in general, considerable mathematical difficulty. However, when the force involves only one variable t, x or v, the integration of the equation is straight forward as we shall presently see.

Case I: Force depends on t only

From Eq. (6.4), we get

$$\frac{\mathrm{d}v}{\mathrm{d}t} = f(t), \tag{6.7}$$

now is of separable type, and we have the first integral

$$\frac{\mathrm{d}x}{\mathrm{d}t} = v = \int f(t)\,\mathrm{d}t + \mathrm{A} = g(t) + \mathrm{A}. \tag{6.8}$$

Integrating Eq. (6.8), we obtain

$$x = \int g(t)\,\mathrm{d}t + \mathrm{A}t + \mathrm{B}. \tag{6.9}$$

In the case of constant value of f Eqs. (6.8) and (6.9) provide the well known Newton's formula for uniform acceleration

$$v = v_0 + ft$$

$$x = v_0\, t + \left(\frac{1}{2}\right) ft^2. \tag{6.10}$$

Case II: Force is a function of x alone

In this case Eq. (6.4) takes the form

$$\frac{\mathrm{d}^2 x}{\mathrm{d}t^2} = f(x) \tag{6.11}$$

Multiplying Eq. (6.11) by 2 $\frac{dx}{dt}$ and integrating, we obtain

$$\left(\frac{dx}{dt}\right)^2 = v^2 = 2\int f(x)\,dx + A \tag{6.12}$$

which on extracting the root gives

$$\frac{dx}{dt} = \pm\sqrt{\left[2\int f(x)\,dx + A\right]}. \tag{6.13}$$

Integration of (6.13) yields

$$t = \pm\int \frac{dx}{\sqrt{\left[2\int^{x} f(y)\,dy + A\right]}} + B. \tag{6.14}$$

Alternately, using Eq. (6.2), we can write Eq. (6.11) as

$$\left(\frac{1}{2}\right)\frac{dv^2}{dx} = f(x) \tag{6.14a}$$

which on integration yields Eq. (6.13).
Remarks: Eq. (6.12) may be expressed as

$$v^2 - v_0^2 = 2\int_{x_0}^{x} f(x)\,dx$$

or

$$\left(\frac{1}{2}\right)m\,(v^2 - v_0^2) = \int_{x_0}^{x} f(x)\,dx, \tag{6.15}$$

and is, thus, seen to represent the work-energy principle for rectilinear motion.

The expression on the right hand side of Eq. (6.13) is assigned the proper +ve or –ve sign, obtained by considering the sign of the increment in x following an increment in t.

Case III: Force depends on v only

In this case equation of motion may be written as

$$\frac{dv}{dt} = f(v). \tag{6.16}$$

Separating the variables and integrating, we get

$$t = \int \frac{dv}{f(v)} + A. \tag{6.17}$$

The above equation may be solved to give

$$v = \frac{dx}{dt} = u(t, A) \tag{6.18}$$

which on integration gives

$$x = \int u(t, \mathrm{A})\, \mathrm{d}t + \mathrm{B}. \tag{6.19}$$

Alternately, Eq. (6.16) may be written as

$$\left(\frac{1}{2}\right)\frac{\mathrm{d}v^2}{\mathrm{d}x} = f(v) \tag{6.20}$$

which after integration and simplification gives

$$\frac{\mathrm{d}x}{\mathrm{d}t} = v = w(x, \mathrm{A}),$$

and then the next integration gives

$$t = \int \frac{\mathrm{d}x}{w(x, \mathrm{A})} + \mathrm{B}. \tag{6.21}$$

These are not the only situations where we can integrate the differential equation of motion in simple terms; there are other special situations also where explicit integration is possible.

In the remaining part of this chapter we shall solve the equation of motion for some specific force functions. However, since the student is already familiar with the case of constant acceleration, we shall not dwell on it.

6.3 SIMPLE HARMONIC MOTION

Simple harmonic motion (SHM) is the simplest of the periodic processes. In a periodic process, a phenomena or event repeats itself periodically at regular intervals. If $f(t)$ analytically represents the phenomena, then it is periodic with time period T, if $f(t + T) = f(t)$. Nature abounds in periodic phenomena. To cite a few examples, mention may be made of motion of planets round the Sun, the motion of earth's artificial satellites, alternation of day and night, vibrations of machines, oscillations in an electric circuit and swinging of a pendulum. The last named is an example of an SHM for small displacement.

A particle is said to execute SHM about a point O, when

(i) it moves on a straight line passing through the centre of force O
(ii) the external force at any instant is proportional to the displacement x of the particle from O, and
(iii) the external force is always directed towards O.

Using Newton's second law and the above definition, the SHM for the particle P is seen to be governed by the second order ordinary differential equation as

$$\frac{\mathrm{d}^2 x}{\mathrm{d}t^2} = -\mu x, \tag{6.22}$$

where $\mu > 0$, thereby ensuring that the force is always directed towards O.

Earlier, we have defined SHM as the motion of a particle moving in a straight line, but more generally, it is the phenomena governed by the simple harmonic equation

$$\frac{d^2 s}{dt^2} = -\mu s, (\mu > O).$$

Solution of the Eq.(6.22) may be attempted by the method mentioned in Section 6.2. Alternatively, the student, familiar with the methods of differential equations, can recognize the equation as a second order equation with constant coefficients having its general solution as

$$x = A \cos \sqrt{\mu}\, t + B \sin \sqrt{\mu}\, t. \tag{6.23}$$

Differentiating above equation with respect to t, we get

$$v = \frac{dx}{dt} = \sqrt{\mu}\, (-A \sin \sqrt{\mu}\, t + B \cos \sqrt{\mu}\, t). \tag{6.24}$$

The constants A and B are evaluated by using the initial conditions. Since only $\cos \sqrt{\mu}\, t$ and $\sin \sqrt{\mu}\, t$ are involved in the expression, we infer that the solution is periodic and bounded. The maximum displacement $|x|$ is called the *amplitude*. Naturally the velocity v vanishes at the positions $|x| = a$ of maximum displacement as the particle has to turn back before it continues with its motion and oscillates in between a and $-a$. Thus, we may write the initial condition as

$$x = a, \frac{dx}{dt} = 0 \text{ at } t = 0. \tag{6.25}$$

Using these in Eqs. (6.23) and (6.24), we get

$$A = a, B = 0. \tag{6.26}$$

Substituting back the values of A and B in Eqs. (6.23) and (6.24), we have

$$x = a \cos \sqrt{\mu}\, t \tag{6.27}$$

$$v = -a\sqrt{\mu} \sin \sqrt{\mu}\, t. \tag{6.28}$$

As both x and v retain the same values after a time $\frac{2\pi}{\sqrt{\mu}}$, the time period of the periodic motion is $\frac{2\pi}{\sqrt{\mu}}$. In other words after a time $\frac{2\pi}{\sqrt{\mu}}$ the particle is at the same position and moving in the same direction with same speed.

Eliminating t between Eqs. (6.27) and (6.28), we obtain

$$v^2 = \mu(a^2 - x^2) \tag{6.29}$$

which also emerges as a first integral of Eq. (6.22).

The general expression Eqs. (6.23) and (6.24) for displacement and speed may also be written as

$$x = a \cos (\omega t + \alpha) \tag{6.30}$$

$$v = -a\omega \sin (\omega t + \alpha). \tag{6.31}$$

where $\omega = \sqrt{\mu}$ is the angular frequency, a the amplitude and $\omega t + \alpha$ the phase at time t, Also the frequency is $\frac{\omega}{2\pi}$.

6.4 SHM UNDER ELASTIC FORCES

A spring is a helix of wire; it can undergo extension as well as compression. A string is normally a strand of rubber or a thin wire; it can only be extended and cannot withstand compression. Under elastic limits when a spring is extended from its natural length a to an extended length x, an internal restoring force of tension T is generated and is given by Hooke's law

$$T = k(x - a), \tag{6.32}$$

where k is the stiffness of the spring.

In the case of elastic string the restoring force of tension is written down by Hooke's law as

$$T = \left(\frac{\lambda}{a}\right)(x - a), \tag{6.33}$$

where λ is the modulus of elasticity.

Since Eqs. (6.32) and (6.33) represent linear restoring forces (i.e. opposing the motion), it is not difficult to comprehend that the ensuing motion under these forces will be simple harmonic in character. Let us study it in more detail.

Let an elastic string AB be of natural length a hanging vertically with the end A tied to a fixed point A and the other end B carrying a particle of mass m. The particle is pulled vertically downward to the point C so that the extension is b (supposed to be within elastic limit) and then released. Now, we will determine the periodic time of motion.

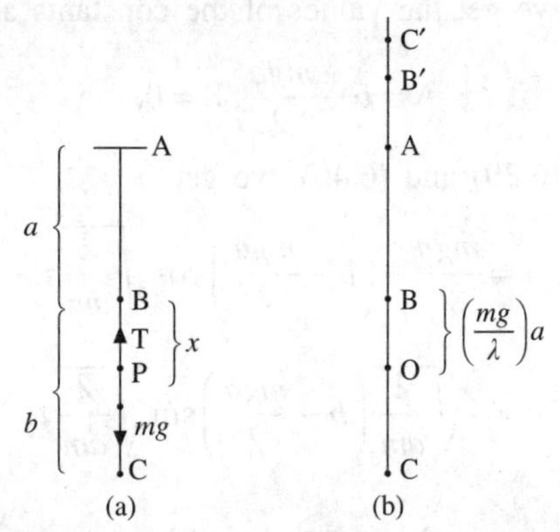

Figure 6.1 Elastic string (a) extended to C showing P as the position of particle at any instant (b) showing the position O of static stretching and the maximum height position C′.

Suppose the extension x (measured downward from B) at any instant of time t is x [refer Figure 6.1]. The forces acting on the particle are its weight mg acting vertically downwards and the tension T acting upwards. From Hooke's law

$$T = \frac{\lambda x}{a}. \tag{6.34}$$

The equation of motion of the particle is given by

$$\frac{d^2 x}{dt^2} = mg - T = -\frac{\lambda}{a}\left(x - \frac{mga}{\lambda}\right) \tag{6.35}$$

or

$$\frac{d^2 X}{dt^2} = -\frac{\lambda}{am} X, \tag{6.36}$$

where
$$X = x - \frac{mga}{\lambda}. \tag{6.37}$$

The initial conditions of the problem are

$$x = b, \frac{dx}{dt} = 0, \text{ at } t = 0. \tag{6.38}$$

Equation (6.36) is simple harmonic equation, and hence, we can write its general solution as

$$X = x - \frac{mga}{\lambda} = \text{A} \cos\sqrt{\frac{\lambda}{am}}\, t + \text{B} \sin\sqrt{\frac{\lambda}{am}}\, t, \tag{6.39}$$

giving

$$\frac{dx}{dt} = v = \sqrt{\frac{\lambda}{am}}\left[-\text{A} \sin\sqrt{\frac{\lambda}{am}}\, t + \text{B} \cos\sqrt{\frac{\lambda}{am}}\, t\right]. \tag{6.40}$$

By using Eq. (6.38) we get the values of the constants as

$$\text{A} = b - \frac{mga}{\lambda}, \text{B} = 0, \tag{6.41}$$

and thus, from Eqs. (6.39) and (6.40), we get

$$x = \frac{mga}{\lambda} + \left(b - \frac{mga}{\lambda}\right) \cos\sqrt{\frac{\lambda}{am}}\, t. \tag{6.42}$$

$$v = -\sqrt{\frac{\lambda}{am}}\left(b - \frac{mga}{\lambda}\right) \sin\sqrt{\frac{\lambda}{am}}\, t. \tag{6.43}$$

Now, in the position ($x = x_0$) of static equilibrium, $\frac{d^2 x}{dt^2} = 0$, and hence it follows that $X = 0$, therefore

$$x_0 = \frac{mga}{\lambda}. \tag{6.44}$$

Thus, the particle executes SHM of period $2\pi\sqrt{\frac{am}{\lambda}}$ about the point O as the centre of force, provided the string at no stage becomes slack. The string can become slack if during the upward motion the particle overshoots the position B. Keeping in view that the amplitude of the motion is $b-\frac{mga}{\lambda}$ [see Eq. (6.42)], the condition for the motion to remain simple harmonic is

$$b-\frac{mga}{\lambda}\le \mathrm{OB}=\frac{mga}{\lambda}$$

or
$$b\le\frac{2mga}{\lambda}.$$

If $b>\frac{2mga}{\lambda}$, the particle in its upward motion will rise above B, and since the string becomes slack, and then it will move freely under gravity.

The time t_1, taken by the particle to move from C to B is obtained by putting $x = 0$ in Eq. (6.42); thus we have

$$t_1=\sqrt{\frac{am}{\lambda}}\left[\cos^{-1}\left(-\frac{mga}{\lambda\left(b-\frac{mga}{\lambda}\right)}\right)\right]. \tag{6.45}$$

Furthermore, using Eq. (6.45) in Eq. (6.43), the velocity of the particle at B is obtained to be

$$V_B=-\sqrt{\frac{\lambda}{am}\left(b^2-\frac{2mgab}{\lambda}\right)}. \tag{6.46}$$

The –ve sign indicates that the particle is moving upwards. Also, the maximum height attained by the particle above B is

$$h=\frac{V_B^2}{2g}=\frac{\lambda}{2mga}\left(b^2-\frac{2mgab}{\lambda}\right), \tag{6.47}$$

provided $h < 2a$, and the time taken (from B) by the particle to attain this height is

$$t_2=\frac{|V_B|}{g}. \tag{6.48}$$

If $h > 2a$, the string will again get stretched, and the force of tension will come into play; this condition may be put in the form

$$\frac{b}{a}>\frac{mg}{\lambda}+\sqrt{\frac{m^2g^2}{\lambda^2}+\frac{4mg}{\lambda}}, \tag{6.49}$$

where account has been taken the fact that $b>\frac{2mga}{\lambda}$

In this case the upward speed with which the particle reaches B′, where AB′ = natural length of the string = a, is

$$|V'_B| = \sqrt{[V_B^2 - 4ga]} = \sqrt{\frac{\lambda}{am}\left[b^2 - \frac{2mga\,(b+2a)}{\lambda}\right]}, \tag{6.50}$$

and the time taken by the particle to move from B to B′ is

$$t'_2 = \frac{\left[|V_B| - |V'_B|\right]}{g}$$

$$= \left[\sqrt{\frac{\lambda}{am}\left(b^2 - \frac{2mgab}{\lambda}\right)} - \sqrt{\frac{\lambda}{am}}\left[b^2 - \frac{2mga\,(b+2a)}{\lambda}\right]\right]. \tag{6.51}$$

Above B′, the motion of the particle is governed by the equation

$$m\frac{\mathrm{d}^2 y}{\mathrm{d}t^2} = -mg - \frac{\lambda y}{a} \tag{6.52}$$

y (> 0) being the upward displacement from B′. Eq.(6.52) may be put in the form

$$\frac{\mathrm{d}^2 Y}{\mathrm{d}t^2} = -\left(\frac{\lambda}{am}\right)Y, \tag{6.53}$$

where

$$Y = y + \frac{mga}{\lambda}. \tag{6.54}$$

Equation (6.53) implies that the motion above B′ is again simple harmonic, and we have

$$y = Y - \frac{mga}{\lambda} = \mathrm{C}\cos\sqrt{\frac{\lambda}{am}}\,t + \mathrm{D}\sin\sqrt{\frac{\lambda}{am}}\,t - \frac{mga}{\lambda} \tag{6.55}$$

and

$$\frac{\mathrm{d}y}{\mathrm{d}t} = \sqrt{\frac{\lambda}{am}}\left[-\mathrm{C}\sin\sqrt{\frac{\lambda}{am}}\,t + \mathrm{D}\cos\sqrt{\frac{\lambda}{am}}\,t\right]. \tag{6.56}$$

Now, counting time from the position of the particle at B′, and applying the conditions

$$y = 0, \frac{\mathrm{d}y}{\mathrm{d}t} = |V'_B| \text{ at } t = 0, \tag{6.57}$$

we get

$$\mathrm{C} = \frac{mga}{\lambda}, \mathrm{D} = \sqrt{\left\{b^2 - \left(\frac{2mga}{\lambda}\right)(b+2a)\right\}}. \tag{6.58}$$

Thus, from Eqs. (6.55) and (6.56), we have

$$y = \sqrt{\left\{b^2 - \left(\frac{2mga}{\lambda}\right)(b+2a)\right\}} \sin \sqrt{\frac{\lambda}{am}}\, t - \left(\frac{mga}{\lambda}\right)\left\{1 - \cos \sqrt{\frac{\lambda}{am}}\, t\right\} \tag{6.59}$$

and

$$\frac{\mathrm{d}y}{\mathrm{d}t} = \sqrt{\frac{\lambda}{am}}\left[\sqrt{\left\{b^2 - \left(\frac{2mga}{\lambda}\right)(b+2a)\right\}} \cos \sqrt{\frac{\lambda}{am}}\, t - \left(\frac{mga}{\lambda}\right) \sin \sqrt{\frac{\lambda}{am}}\, t\right] \tag{6.60}$$

The particle will come to rest at the point C′ after a time t_3 given by

$$\left(\frac{mga}{\lambda}\right) \tan \sqrt{\frac{\lambda}{am}}\, t_3 = \sqrt{\left\{b^2 - \left(\frac{2mga}{\lambda}\right)(b+2a)\right\}}, \tag{6.61}$$

and the height of C′ above B′ is

$$h' = y\big|_{t=t_3} = \sqrt{\left[\left\{b^2 - \left(\frac{2mga}{\lambda}\right)(b+2a)\right\} + \frac{m^2 g^2 a^2}{\lambda^2}\right]} - \frac{mga}{\lambda}.$$

It may be noted that h' is positive because of Eq. (6.49). Further Eqs. (6.53) and (6.54) show that the centre of force is located at $y = -\dfrac{mga}{\lambda}$, i.e. at a depth $\dfrac{mga}{\lambda}$ below B′ (although there is no SHM for $y < 0$). This may be interpreted (with B′ merged into B) that the effective centre of force is still the position O of static equilibrium.

Since the gravitational force and the elastic force are both conservative, the particle will return to the starting point C with zero velocity in each of the three cases. The motion is periodic and the time taken by the particle to return to C is given in the three cases as follows:

Case I: When $b \le 2\dfrac{mga}{\lambda}$

$$T_1 = 2\pi \sqrt{\frac{am}{\lambda}}.$$

Case II: When $2\dfrac{mga}{\lambda} < b < a\left[\dfrac{mg}{\lambda} + \sqrt{\dfrac{m^2g^2}{\lambda^2} + \dfrac{4mg}{\lambda}}\right]$

$$T_2 = 2t_1 + 2t_2,$$

where t_1 and t_2 are given by Eqs. (6.45) and (6.48)

Case III: When $b > a\left[\frac{mg}{\lambda} + \left\{\frac{mg}{\lambda} + \sqrt{\left(\frac{m^2g^2}{\lambda^2} + \frac{4mg}{\lambda}\right)}\right\}\right]$

$$T_3 = 2t_1 + 2t_2' + 2t_3,$$

where t_1, t_2' and t_3 are given by Eqs. (6.45), (6.51) and (6.61).

6.5 MOTION UNDER INVERSE SQUARE LAW

According to Newton's gravitational law a mass attracts with a force inversely proportional to the square of the distance from its centre of mass. Considering the earth to be a sphere of radius a of uniform mass density, the force of attraction on a particle, e.g., a falling raindrop or a soaring rocket, at a distance $x(> a)$ from its centre may be written as $\frac{\mu}{x^2}$. Moreover, since the value of gravitational force on earth's surface $x = a$ is g, we may write $\mu = ga^2$. Thus, for exterior motion, when the external force is gravitational force only, the equation of motion is given by

$$\frac{d^2x}{dt^2} = -\frac{ga^2}{x^2}, (x > a). \tag{6.62}$$

This corresponds to Case II of Section 6.2.
Thus, on integrating Eq. (6.62) yields

$$\left(\frac{dx}{dt}\right)^2 = \frac{2ga^2}{x} + \text{A}. \tag{6.63}$$

Case I: Motion towards earth's surface

The particle is assumed to start at a height h above the earth's surface, the initial conditions are

$$x = a + h, \frac{dx}{dt} = 0 \text{ at } t = 0. \tag{6.64}$$

Using the second of above conditions (6.64) in (6.63), we get

$$\text{A} = -\frac{2g^2a}{(a+h)},$$

and hence Eq. (6.63) provides

$$v = \frac{dx}{dt} = -a\sqrt{2g\left(\frac{1}{x} - \frac{1}{a+h}\right)} \tag{6.65}$$

–ve sign is affixed because x decreases with t. Separating the variables, integrating and using first conditions given in Eq. (6.64), Eq. (6.65) yields

$$t = \sqrt{\frac{a+h}{2g}}\left[\frac{a+h}{a}\cos^{-1}\sqrt{\frac{x}{a+h}} + \frac{\sqrt{\{x(a+h-x)\}}}{a}\right]. \tag{6.66}$$

This gives the time to reach the surface as

$$t_a = t\,|_{x=a} = \sqrt{\frac{a+h}{2g}}\left[\frac{a+h}{a}\sin^{-1}\sqrt{\frac{h}{a+h}} + \sqrt{\frac{h}{a}}\right]. \tag{6.67}$$

Case II: Motion away from earth's surface

Let the particle be projected vertically upwards with initial velocity u from earth's surface. In this case the initial conditions are

$$v = \frac{\mathrm{d}x}{\mathrm{d}t} = u,\ x = a,\ \text{at } t = 0. \tag{6.68}$$

Applying conditions given in Eq. (6.68) to Eq. (6.63), we get

$$u^2 = 2ga + \mathrm{A}. \tag{6.69}$$

Eliminating A in between Eqs. (6.63) and (6.69) and taking their square roots, we get

$$\frac{\mathrm{d}x}{\mathrm{d}t} = \sqrt{\frac{(u^2 - 2ga)(c+x)}{x}},\quad c = \frac{2ga^2}{u^2 - 2ga} \tag{6.70}$$

+ve sign being assigned as x increases with t.

Now, separating the variables and integrating, we get the time to reach a distance x from the earth's surface as

$$t = \frac{\sqrt{x(c+x)} - \sqrt{a(c+a)} - c\log(\sqrt{x} + \sqrt{c+a}) + c\log(\sqrt{a} + \sqrt{c+a})}{\sqrt{(u^2 - 2ga)}}. \tag{6.71}$$

Case III: Motion inside the earth

As calculated from Newton's gravitational law, the force of attraction, inside the earth, varies as the distance from the centre. Therefore, since the value of the gravitational force at a surface point is g, the equation of motion for a particle moving in a vertical shaft towards the centre of the earth is

$$\frac{\mathrm{d}^2 x}{\mathrm{d}t^2} = -\frac{g}{a}x(x \le a). \tag{6.72}$$

Integrating Eq. (6.65), we get

$$v^2 = \left(\frac{\mathrm{d}x}{\mathrm{d}t}\right)^2 = \mathrm{B} - \frac{gx^2}{a}. \tag{6.73}$$

If the motion is a continuation of the motion considered above, we obtain from Eq. (6.65) the condition

$$v^2 = \frac{2gah}{a+h} \text{ at } x = a. \tag{6.74}$$

Using this value in Eq. (6.73), we get

$$B = \frac{ga(3h+a)}{(a+h)},$$

which on being fed back into Eq. (6.73) leads to

$$\frac{dx}{dt} = v = \sqrt{\frac{g}{a}\left(\frac{3h+a}{h+a}a^2 - x^2\right)}. \tag{6.75}$$

Separating the variables and integrating, we get

$$t = t_a - \int_a^x \frac{dy}{\sqrt{\frac{g}{a}\left(\frac{3h+a}{h+a}a^2 - y^2\right)}} \tag{6.76}$$

$$= t_a + \sqrt{\frac{a}{g}}\left[\sin^{-1}\sqrt{\frac{h+a}{3h+a}} - \sin^{-1}\frac{x}{a}\sqrt{\frac{h+a}{3h+a}}\right],$$

where we have used the condition $x = a$, when $t = t_a$, t being given by Eq. (6.67).

Next, taking $x = 0$ in Eq. (6.69), we obtain the time t_0 to reach the centre of the earth as

$$t_0 = t_a + \sqrt{\frac{a}{g}} \sin^{-1}\sqrt{\frac{h+a}{3h+a}}. \tag{6.77}$$

Obviously the time of a complete oscillation is

$$T = 4t_0 = \frac{4}{a\sqrt{2g}}\left[(a+h)^{\frac{3}{2}} \sin^{-1}\sqrt{\frac{h}{h+a}} + \sqrt{2}a^{\frac{3}{2}} \sin^{-1}\sqrt{\left(\frac{h}{3h+a}\right)} + \sqrt{ah(a+h)}\right]. \tag{6.78}$$

Also, the speed of the particle on reaching the centre of the earth is

$$v = \left.\frac{dx}{dt}\right|_{x=0} = \sqrt{\frac{ga(3h+a)}{(h+a)}}. \tag{6.79}$$

It may be noted that in terms of the gravitational constant G and earth's mass M, we have

$$g = \frac{GM}{a^2}.$$

The motion under an external force given as a function of x may be studied in a similar way.

6.6 MOTION IN A RESISTING MEDIUM

A rocket fired into earth's atmosphere, a gas bubble rising in an oil chamber, a boat paddling in a lake, a bird flying in air or a micro-organism swimming under water, all experience a force of resistance-naturally called the *Drag*. In this section we shall investigate the effect of drag force on the vertical motion of a particle moving under the uniform acceleration g near the surface of the earth. Besides, depending on the nature of the ambient fluid, the drag $D(v)$, per unit mass, has been found to be an increasing function of the relative speed v of the body. Experiments reveal and theoretical considerations predict that there are different resistance laws depending on the range of the speed of the particle. For motion at ordinary speeds in the air, the resistance is found to be proportional to the square of the speed and for slow motion in viscous fluids it is proportional to the speed; there are other laws too.

The equation of motion for a particle moving vertically downwards is expressible as

$$\frac{d^2 x}{dt^2} = \frac{1}{2}\frac{dv^2}{dx} = g - D(v), \tag{6.80}$$

where x is also being measured downwards. It is obvious from Eq. (6.80) that at certain velocity $v = V$, the acceleration vanishes so that the subsequent motion is uniform. Therefore, V is called *terminal velocity* or *critical velocity*. Thus, critical velocity is given by

$$D(V) = g. \tag{6.81}$$

The term critical velocity is retained even when the motion is upwards. We shall demonstrate the effect of the drag force further.

6.6.1 Resistance Proportional to the Particle's Speed

Let a particle be thrown vertically upwards with speed U in a medium for which the resistance is proportional to the particle's speed. The particle's terminal speed is V. Let us now learn to calculate

(i) the length of the interval during which the particle is moving upwards
(ii) the height that the particle attains
(iii) the amount of energy dissipated during the upward motion
(iv) the speed with which the particle returns to the point of projection
(v) the time in which the particle returns to the point of projection

The entire motion may be broken into two parts.

Upward motion

Let at any instant of time the particle be at P, a height x vertically above the point of projection O. The particle experiences a force mg, due to the force of gravity and a drag force mkv, both acting downwards; here m is the mass of the particle and v its velocity.

Clearly from the condition given in Eq. (6.81), we have

$$g = kV, \tag{6.82}$$

and hence equation of motion may be written as

$$\frac{\mathrm{d}v}{\mathrm{d}t} = -k(V+v). \tag{6.83}$$

Separating the variables and integrating above equation, we get

$$V + v = \mathrm{A}\, e^{-kt},$$

and then the initial condition, $v = U$ at $t = 0$, provides the value $\mathrm{A} = V + U$; thus, we have

$$v = \frac{\mathrm{d}x}{\mathrm{d}t} = (U+V)\, e^{-kt} - V. \tag{6.84}$$

Now, the particle will rise for a time t_1 till its velocity vanishes; hence we have from Eq. (6.84)

$$t_1 = -\frac{1}{k}\log\left(1+\frac{U}{V}\right) = \frac{V}{g}\log\left(1+\frac{U}{V}\right). \tag{6.85}$$

Further integrating Eq. (6.84) and using the initial condition $x = 0$ at $t = 0$, we get

$$x = \frac{1}{k}(U+V)(1-e^{-kt}) - Vt. \tag{6.86}$$

Putting $t = t_1$ from Eq. (6.85) in Eq. (6.86), we obtain the greatest height x_1 attained by the particle

$$x_1 = \frac{U}{k} - \frac{V}{k}\log\left(1+\frac{U}{V}\right) \tag{6.87}$$

$$= \frac{V}{g}\left[U - V\log\left(1+\frac{U}{V}\right)\right]. \tag{6.88}$$

Next, the amount of energy dissipated during the upward motion is

$$\text{Loss of K.E.} - \text{Gain in P.E.} = \frac{1}{2}mU^2 - mgx_1 \tag{6.89}$$

$$= \frac{1}{2}mU^2 - mV\left[U - V\log\left(1+\frac{U}{V}\right)\right]. \tag{6.90}$$

Downward motion

Here it will be convenient to place our origin O′ at the highest point attained by the particle (OO′ = x_1) and start counting time again from its position at O′. Let the particle be at a depth x below O′ at time t. Now, the force of gravity g acts downwards and the drag force kv upwards, and hence, the equation of motion is

$$v\frac{\mathrm{d}v}{\mathrm{d}x} = k(V - v). \tag{6.91}$$

Integrating Eq. (6.91) and using the condition $v = 0$ at $x = 0$, we obtain

$$x = \frac{V}{k}\log\frac{V}{V-v} - \frac{v}{k} = \frac{V^2}{g}\log\frac{V}{V-v} - \frac{Vv}{g}. \tag{6.92}$$

The speed W with which the particle finally returns to the point of projection is obtained by taking $x = x_1$ [as given by Eq. (6.87)] in Eq. (6.92); thus, we have

$$W = V\log\frac{V+U}{V-W} - U. \tag{6.93}$$

The equation of motion may also be expressed as

$$\frac{\mathrm{d}v}{\mathrm{d}t} = k(V - v), \tag{6.94}$$

which on integration, under the condition $v = 0$ at $t = 0$, yields

$$v = V(1 - e^{-kt}). \tag{6.95}$$

Now, the time t_2 in which the particle moves from O′ to O is obtained from Eq. (6.95) by taking $v = W$; thus, we have

$$t_2 = \frac{V}{g}\log\frac{V}{V-W}. \tag{6.96}$$

Hence, the total time T in which the particle returns to the point of projection is given by

$$T = t_1 + t_2$$

or

$$T = \frac{V}{g}\log\frac{V+U}{V-W} = \frac{W+U}{g}. \tag{6.97}$$

6.6.2 Resistance Proportional to the Square of the Speed

Let us now consider the case when the resistance is proportional to the square of the speed.

Upward motion

In this case the equation of motion may be written as

$$v\frac{\mathrm{d}v}{\mathrm{d}x} = k(V^2 - v^2), \tag{6.98}$$

where now $k = \dfrac{g}{V^2}$.

Integrating Eq. (6.98), under the condition $v = U$ when $x = 0$, we get

$$x = \frac{1}{2k} \log \frac{V^2 + U^2}{V^2 + v^2}, \tag{6.99}$$

giving

$$v^2 = (V^2 + U^2)\, e^{-2kx} - V^2. \tag{6.100}$$

Therefore,

$$\frac{\mathrm{d}x}{\mathrm{d}t} = v = -V\, e^{-kx} \sqrt{\left[\frac{V^2 + U^2}{V^2} - e^{2kx}\right]}. \tag{6.101}$$

Integrating Eq. (6.94), under the condition $x = 0$ at $t = 0$, we get

$$kVt = \sin^{-1}\left[\frac{V}{\sqrt{V^2 + U^2}}\, e^{kx}\right] - \tan - \frac{V}{U} \tag{6.102}$$

or

$$x = \frac{1}{k} \log\left[\sqrt{\left(1 + \frac{U^2}{V^2}\right)} \sin\left(kVt + \tan^{-1}\frac{V}{U}\right)\right].$$

We can eliminate x in between Eqs. (6.99) and (6.102) to get the relation.

$$kVt = \tan^{-1}\frac{V}{v} - \tan^{-1}\frac{V}{U} = \tan^{-1}\frac{U}{V} - \tan^{-1}\frac{v}{V}. \tag{6.103}$$

We could have arrived at this equation by taking t to be the independent variable as was done in the Section 6.6.1.

The particle will cease to move upwards when $v = 0$; thus, Eq. (6.103) provides the required time t as

$$t_1 = \frac{1}{Vk} \tan\frac{U}{V}. \tag{6.104}$$

Again, putting $v = 0$ in Eq. (6.99), we get the maximum height x_1 gained by the particle as

$$x_1 = \frac{1}{2k} \log\left(1 + \frac{U^2}{V}\right). \tag{6.105}$$

The amount of energy dissipated during upward motion is

$$E = \frac{1}{2} mU^2 - mgx_1 = \frac{1}{2} m\left[U^2 - V^2 \log\left(1 + \frac{U^2}{V^2}\right)\right] \tag{6.106}$$

Downward motion

With x now being measured downwards from the highest point, the equation of motion may be written as

$$\frac{1}{2}\frac{\mathrm{d}v^2}{\mathrm{d}x} = k\,(V^2 - v^2). \tag{6.107}$$

Integrating Eq. (6.107), under the condition $v = 0$ when $x = 0$, we obtain

$$v^2 = V^2\,(1 - e^{-2kx}), \tag{6.108}$$

providing

$$\frac{\mathrm{d}x}{\mathrm{d}t} = v = V\sqrt{(1 - e^{-2kx})}. \tag{6.109}$$

Integrating Eq. (6.109) under the condition $x = 0$ when $t = 0$, we get

$$x = \frac{1}{k}\log\cosh kVt. \tag{6.110}$$

Putting $x = x_1$ as given by Eq. (6.105) in Eq. (6.108), we get the velocity W with which the particle returns to the point of projection; thus, we have

$$W = \frac{VU}{\sqrt{V^2 + U^2}}. \tag{6.111}$$

Further, putting $x = x_1$ in Eq. (6.110), we obtain time t_2 in which the particle moves from the highest point to the initial point; thus, we have

$$t_2 = \frac{1}{kV}\log\left[\sqrt{\left(1 + \frac{U^2}{v^2}\right)} + \frac{U}{V}\right]. \tag{6.112}$$

Now, the total time T is which the particle returns to the point of projection follows from Eqs. (6.104) and (6.112) as

$$T = t_1 + t_2. \tag{6.113}$$

6.7 MOTION OF A ROCKET

A rocket essentially consists of the mass of unburnt fuel (solid or liquid propellant), the mass of structure (casing, engine, etc.) and the mass of payload (a warhead, a satellite to be put into orbit). It works on Newton's third law. Through chemical reaction the fuel gets converted into a gas which escapes from the tail of the rocket at a very high speed, thereby producing a reactive force in the opposite direction. The main difference between a rocket engine and a jet engine is that while the latter is air breathing the former requires no atmosphere and actually works more efficiently in the outer space. Since the fuel is being emitted from the rocket, the problem is that of varying mass, the basic equation for which we now establish.

Suppose a body of mass $m(t)$ at time t is moving with velocity $\mathbf{u}$. Let mass be ejected from it at a rate α and with velocity $\mathbf{v}$ relative to the body. Thus at time $t + \delta t$, the body will be having a mass $m - \alpha\delta t$ and moving with velocity $\mathbf{u} + \delta\mathbf{u}$. On the other hand, the ejected mass $\alpha\delta t$ will be moving with velocity $\mathbf{u} + \mathbf{v}$. Therefore, the change of momentum in time δt is

$$(m - \alpha\delta t)(\mathbf{u} + \delta\mathbf{u}) + \alpha\delta t(\mathbf{u} + \mathbf{v}) - m\mathbf{u} = m\,\delta\mathbf{u} + \alpha\mathbf{v}\,\delta t,$$

where only first order terms have been retained.
Now, the rate of change of momentum is

$$\lim_{\delta t \to 0}\left[m\frac{\delta\mathbf{u}}{\delta t} + \alpha\mathbf{v}\right] = m\frac{\mathrm{d}\mathbf{u}}{\mathrm{d}t} + \alpha\mathbf{v}.$$

If $\mathbf{F}$ is the external body force, then we have from Newton's second law

$$m\frac{\mathrm{d}\mathbf{u}}{\mathrm{d}t} + \alpha\mathbf{v} = \mathbf{F}$$

or

$$m\frac{\mathrm{d}\mathbf{u}}{\mathrm{d}t} = \mathbf{F} - \alpha\mathbf{v}. \tag{6.114}$$

In the case of rocket motion α is positive, but $\mathbf{v}$ (being opposite to $\mathbf{u}$) is negative and so the reactive force $-\alpha\mathbf{v}$ is positive, i.e. it is aiding the increase of rocket velocity $\mathbf{u}$.

Here, we shall be concerned with vertical motion (k-direction) of a rocket and hence, we may set

$$\mathbf{u} = u\hat{\mathbf{k}},\quad \mathbf{v} = -v\hat{\mathbf{k}},\quad \mathbf{F} = -g\hat{\mathbf{k}},\quad m(t) = M + P - \alpha t,$$

where M is the initial mass of rocket (casing, engine, unburnt fuel) and P is the payload. A parameter of importance is the time T of attaining all burnt stage, i.e. when all fuel has been exhausted; beyond time T the reactive force will cease to act. Thus, Eq. (6.114) assumes the following form for the rectilinear motion of a rocket

$$(M + P - \alpha t)\frac{\mathrm{d}u}{\mathrm{d}t} = -(M + P - \alpha t)g + \alpha v. \tag{6.115}$$

If the initial velocity is u_0 the solution of Eq. (6.115) is given by

$$u = u_0 - v\log\left[1 - \frac{\alpha t}{M + P}\right] - gt\ (0 \le t \le T). \tag{6.116}$$

The time T in practical cases is so small that the gravitational term is negligible in comparison to the second term at least at the all burnt stage $t = T$. Thus, taking α to be large the ultimate speed U may be expressed as

$$U = v\log\frac{P + M}{P + M - \beta M}, \tag{6.117}$$

when the initial speed $u_0 = 0$ and where $\beta = \dfrac{\alpha T}{M}$ is the ratio of the total mass of the fuel to the mass M.
Writing $P + M = M_i$, the total initial mass, and $M_i - \alpha T = M_f$, the final mass at all burnt stage, Eq. (6.117) leads to

$$M_i = M_f \, e^{\frac{U}{v}}. \tag{6.118}$$

Above relation is known as *Tsiolkovsky's formula*. It demonstrates the fact that in order to attain high values of U with some significant final mass M_f the initial mass M_i has to be enormously large. It may be shown that to attain speed beyond 10 km/sec technical difficulties in the construction of the rocket arise.

This becomes all the more significant when we consider the fact that the escape velocity is approximately 11.2 km/sec. In order to overcome this difficulty, multistage rockets are constructed. Two stage rocket is designed in such a way that when the fuel in the first stage is burnt out, the first stage casing, etc., falls off, and the second stage engine takes over. With P as the payload let us suppose that M_1 is the mass of the first stage and M_2 the mass of the second stage. Further, assume that βM_1 is the initial mass of the fuel in the first stage and βM_2 in the second stage.

Exploiting Eq. (6.117), we get the speed U_1 attained by the rocket when the first stage is working as

$$U_1 = v \log \frac{P + M_2 + M_1}{P + M_2 + M_1'}, \tag{6.119}$$

where $M_1' = (1 - \beta)\, M_1$ is the mass of the first stage without the fuel.

The mass M_1' is cast off and taking the initial velocity now as U_1 [see Eq. (6.116)] we can write the final velocity at the end of second stage

$$U_2 = U_1 + v \log \frac{P + M_2}{P + M_2'} \tag{6.120}$$

$$= v \log \frac{(P + M_2)\,(P + M_2 + M_1)}{(P + M_2')\,(P + M_2 + M_1')}, \tag{6.120a}$$

where $M_2' = (1 - \beta)\, M_2$ is the mass of the second stage without the fuel.

Students are advised to show that in a three stage rocket, with obvious meaning of the symbols, Eq. (6.120a) assumes the form

$$U = v \log \frac{(P + M_3)\,(P + M_3 + M_2)\,(P + M_3 + M_2 + M_1)}{(P + M_3')\,(P + M_3 + M_2')\,(P + M_3 + M_2 + M_1')}, \tag{6.121}$$

and this may be continued to an n-stage rocket. To put an artificial satellite in an orbit round the earth it may be shown that a three stage rocket is most suitable.

SOLVED EXAMPLES

EXAMPLE 6.1 A horizontal shelf is moved up and down with SHM of period 1 sec. What is the greatest amplitude admissible in order that a weight placed on the shelf may not be jerked off?

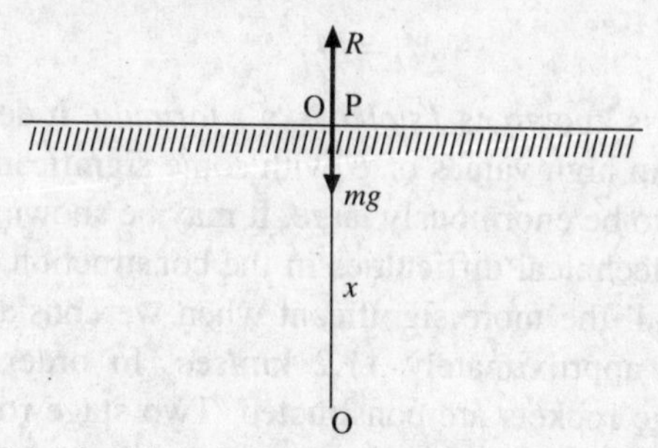

Figure 6.2 Particle P in contact with the shelf executing SHM.

Solution Let the position of the shelf at an instant of time t be at a distance x from the centre of Force O, Since the shelf is executing SHM, we have

$$\frac{d^2 x}{dt^2} = -\mu x. \tag{i}$$

When in contact with the shelf, the particle P of mass m experiences a force of reaction R in addition to the force of its weight mg; thus, its equation of motion is

$$m\frac{d^2 x}{dt^2} = R - mg, \tag{ii}$$

providing on using Eq. (i)

$$\frac{R}{m} = g - \mu x. \tag{iii}$$

Now, in order that the weight does not loose contact with the shelf, we must have $R_{min} > 0$, and hence, from Eq. (iii), this amounts to

$$g - \mu x_{max} > 0. \tag{iv}$$

Observing that x_{max} is the amplitude a, Eq. (iv) reduces to

$$a < \frac{g}{\mu}, \tag{v}$$

and this clearly shows that maximum admissible amplitude in order that the weight is not jerked off the shelf is

$$a_{max} = \frac{g}{\mu} = \frac{g}{4\pi^2},$$

as in the present situation with $T = 1$, we have $\mu = \dfrac{4\pi^2}{T^2} = 4\pi^2$.

EXAMPLE 6.2 A particle describes a circle with uniform speed; show that its projection on a diameter executes a SHM about the centre of the circle.

Solution Since the particle is describing the circle with uniform speed its angular speed

$$\frac{d\theta}{dt} = \omega \tag{i}$$

is a constant, and hence

$$\theta = \omega t. \tag{ii}$$

Now, for the projection Q of the particle P, we have

$$x = \text{OQ} = a \cos \theta = a \cos \omega t, \tag{iii}$$

where a is the radius of the circle.
Differentiating both sides of Eq. (iii), we obtain

$$\frac{d^2 x}{dt^2} = -\omega^2 x,$$

which is simple harmonic equation showing that the particle executes SHM about the centre O.

EXAMPLE 6.3 Show that a particle executing SHM requires one-sixth of its period to move from the position of maximum displacement to one in which the displacement is half the amplitude.

Solution The time period of SHM governed by Eq. (6.22) is

$$T = \frac{2\pi}{\sqrt{\mu}}, \tag{i}$$

and the displacement x, from Eq. (6.27), is

$$x = a \cos \sqrt{\mu} t. \tag{ii}$$

Taking $x = \frac{a}{2}$ in Eq. (ii), it follows that required time is

$$t = \frac{1}{\sqrt{\mu}} \cos^{-1}\left(\frac{1}{2}\right) = \frac{\pi}{3\sqrt{\mu}} = \frac{T}{6}.$$

EXAMPLE 6.4 A particle moves in a straight line with acceleration, towards a fixed point O in the straight line, equal to $\frac{\mu}{x^2} - \frac{\lambda}{x^3}$, when the particle is at a distance x from the point; it starts from rest at a distance a. Determine the periodic time.

Solution Equation of motion is

$$v \frac{dv}{dx} = \frac{d^2 x}{dt^2} = -\frac{\mu}{x^2} + \frac{\lambda}{x^3}. \tag{i}$$

Integrating Eq. (i), we obtain

$$v^2 = \left(\frac{\mathrm{d}x}{\mathrm{d}t}\right)^2 = \frac{2\mu}{x} - \frac{\lambda}{x^2} + \mathrm{A}. \tag{ii}$$

Applying the condition $v = 0$ at $x = a$, we get $\mathrm{A} = \dfrac{\lambda}{a^2} - \dfrac{2\mu}{a}$.

Putting the value in (ii), we have

$$v = \frac{\mathrm{d}x}{\mathrm{d}t} = -\frac{\sqrt{(a-x)\{(2\mu a - \lambda)\,x - \lambda a\}}}{ax}, \tag{iii}$$

where –ve sign is assigned because x decreases with t. Equation (iii) clearly shows that

$$a \geq x \geq \frac{\lambda a}{(2\mu a - \lambda)},$$

i.e., the particle oscillates between the limits a and $\dfrac{\lambda a}{(2\mu a - \lambda)}$.

Further, separating the variables and integrating Equation (iii), we find that the periodic time T is given by

$$T = 2\int_{\frac{\lambda a}{(2\mu a-\lambda)}}^{a} \frac{\mathrm{d}x}{\sqrt{(a-x)\{(2\mu a - \lambda)\,x - \lambda a\}}}$$

$$= \frac{2\pi\mu a^3}{(2\mu a - \lambda)^{\frac{3}{2}}}.$$

EXAMPLE 6.5 A uniform flexible chain of length a rests in a straight line on a smooth horizontal table except for a length b which hangs over an edge at right angles to it. If the chain moves from rest, show that after a time t the length x of the overhanging portion is given by $x = b\cosh\sqrt{\dfrac{g}{a}}\,t$.

Solution Since the weight of the overhanging part is the driving force, the equation of motion is

$$\frac{\mathrm{d}^2 x}{\mathrm{d}t^2} = \frac{g}{a}x. \tag{i}$$

The general solution of the above equation is

$$x = \mathrm{A}\cosh\sqrt{\frac{g}{a}}\,t + \mathrm{B}\sinh\sqrt{\frac{g}{a}}\,t \tag{ii}$$

giving

$$\frac{dx}{dt} = \sqrt{\frac{g}{a}}\left[A \sinh \sqrt{\frac{g}{a}}\, t + B \cosh \sqrt{\frac{g}{a}}\, t\right]. \qquad \text{(iii)}$$

Applying Eqs. (i), (ii) and (iii)

$$x = b, \frac{dx}{dt} = 0 \text{ at } t = 0,$$

we get

$$A = b \text{ and } B = 0,$$

and hence the solution.

PROBLEMS

1. If v_1, v_2 are the velocities of a particle moving in SHM at distances x_1, x_2 from the centre, show that the time of a complete oscillation is

$$2\pi \sqrt{\left[\frac{\left(x_1^2 - x_2^2\right)}{\left(v_2^2 - v_1^2\right)}\right]}.$$

2. A particle of mass m is attached to a light wire which is stretched tightly between two fixed points with a tension T in a smooth horizontal plane. If x_1 and x_2 are the distances of the particle from the fixed points, show that the period of small oscillation of the particle is $2\pi \sqrt{\left[\frac{m x_1 x_2}{(x_1 + x_2)T}\right]}$.

3. A particle of mass m is placed on a horizontal board which is made to execute vertical SHM of period T and amplitude a. If $a < \left(\frac{gT^2}{4\pi}\right)$, show that the particle does not loose contact with the board at any time.

4. A point executes SHM such that in two of its positions the velocities are u_1 and u_2 and the corresponding accelerations are f_1 and f_2. Find the distance between the positions, and the amplitude of the motion.

$$\left[\textbf{Ans: } \frac{u_2^2 \sim u_1^2}{f_1 + f_2};\ \frac{1}{f_2^2 \sim f_1^2}\left[\left(u_2^2 \sim u_1^2\right)\left(f_1^2 u_2^2 \sim f_2^2 u_1^2\right)\right]^{1/2}\right]$$

5. A particle moves in a straight line OCP being attracted by a force μm PC always directed towards C, whilst C moves along OC with constant acceleration f. If initially C was at O, and P was at a distance c from O

and moving with velocity v, prove that the distance of P from O at any time t is

$$\left(\frac{f}{\mu}+c\right)\cos(t\sqrt{\mu})+\frac{v}{\sqrt{\mu}}\sin(t\sqrt{\mu})-\frac{f}{\mu}+\frac{1}{2}ft^2.$$

6. A particle of mass m oscillates in a line with natural period $\frac{2\pi}{n}$. If an applied periodic force F cos pt acts in the line so that the particle is instantaneously at rest at zero time at a distance d from the centre of oscillation, prove that the displacement of the particle at a subsequent time t is

$$d\cos nt+\frac{F(\cos pt-\cos nt)}{(n^2-p^2)m}.$$

If V is the maximum speed attained in the presence of the force $F\cos pt$, then show that the angular frequency of the free oscillations is $\left[\frac{(F+mpV)p}{mV}\right]$.

7. A particle rests in equilibrium under the attraction of two centre of forces which attract directly as the distance, their attractions per unit of mass at unit distances being μ and μ'; the particle is slightly displaced towards one of them, show that the time of a small oscillation is $\frac{2\pi}{\sqrt{(\mu+\mu')}}$.

8. An endless cord consists of two portions, of lengths $2l$ and $2l'$, respectively, knotted together, their masses per unit length being m and m'. It is placed in stable equilibrium over a small smooth peg and then slightly displaced. Show that the time of a complete oscillation is

$$2\pi\sqrt{\left[\frac{ml+m'l'}{(m-m')g}\right]}.$$

9. Two masses m_1 and m_2 are connected by a spring of such a strength that when m_1 is held fixed, m_2 performs n complete oscillations per second. Show that if m_2 be held fixed, m_1 will make $n\sqrt{\left(\frac{m_2}{m_1}\right)}$, and if both be free they will make $n\sqrt{\left[\frac{(m_1+m_2)}{m_1}\right]}$ oscillations per second, the oscillations in each case being in the line of the springs.

10. A heavy particle of mass m is attached to one end of an elastic string of natural length a, whose other end is fixed at O. The particle falls from rest at O. Show that part of the motion is simple harmonic, and that, if the greatest depth of the particle below 0 is $a\cot^2\left(\frac{\theta}{2}\right), \left(0 < \theta < \frac{\pi}{2}\right)$, the modulus of elasticity of the string is $\frac{1}{2} mg \tan^2 \theta$. Further, show that the particle attains this depth in time

$$\left(\frac{2a}{g}\right)^{\frac{1}{2}} [1 + (\pi - \theta) \cot \theta].$$

11. Two bodies of masses m and m' are attached to the lower end of an elastic string whose upper end is fixed, and hang at rest. The mass m' falls off; show that the distance of m from the upper end of the string at time t is

$$a + b + c \cos\left(\sqrt{\frac{g}{b}}\, t\right),$$

where a is the unstretched length of the string, and b and c are the distances by which it would be stretched when supporting m and m' respectively.

12. One end of an elastic string is fastened at A; to the other end is fastened a particle heavy enough to stretch the string to double its natural length a. Show that if the particle is dropped from A, it will descend a distance $(2 + \sqrt{3})\, a$ before coming to rest.

13. A particle P, of unit mass, moves under the action of repulsive force ω^2 OP. The particle is initially projected from a point A, with speed $a\omega\sqrt{6}$, in the direction AOB, where OA = a and OB = $2a$. Show that the time from A to B is $\frac{1}{\omega} \log\left(1 + \sqrt{6}\right)$.

14. A heavy particle is attached to one end of an elastic string the other end of which is fixed. The modulus of elasticity of the string is equal to the weight of the particle. The string is drawn vertically down till it is four times its natural length and is then released. Show that the particle will return to this point in time

$$\sqrt{\left(\frac{a}{g}\right)}\left(\frac{4\pi}{3} + 2\sqrt{3}\right),$$

where a is the natural length of the string.

15. A weightless elastic string, of natural length l and modulus λ, has equal particles of mass m at its ends and lies on a smooth horizontal table perpendicular to an edge with one particle just hanging over. Show that the other particle will pass over at the end of time t given by the equations.

$$2l + \frac{mgl}{\lambda} \sin^2\left(\sqrt{\frac{\lambda}{2ml}}\, t\right) = \frac{1}{2} gt^2.$$

16. A small bead P can slide on a smooth wire AB, being acted upon by a force per unit of mass equal to $\frac{m}{CP^2}$ from a point C which is outside AB. Show that the time of a small oscillation about its mean position is $\left(\frac{2\pi}{\sqrt{\mu}}\right)p^{\frac{3}{2}}$, where p is the perpendicular distance of C from AB.

17. A particle moves along the axis of x starting from rest at $x = a$, for an interval t_1, from the beginning of the motion the acceleration is $-\mu x$; for a subsequent interval t_2, the acceleration is μx, and at the end of this interval the particle is at origin; prove that

$$\tan(t_1\sqrt{\mu})\cdot\tanh(t_2\sqrt{\mu}) = 1.$$

18. Assuming that gravity inside the earth varies distance from its centre, show that a train, starting from rest and moving under gravity only, would take the same time to traverse a smooth straight airless tunnel between any two points of the earth's surface. Find the time.

[**Ans:** 42.5 mts. approx.]

19. A particle P of mass m moves along a straight line through a point O and at any instant the distance OP is x. When $x > a$, the particle is attracted towards O by a force $\frac{m\mu}{x^2}$, and when $x < a$ the particle is repelled from O by a force $\frac{ma\mu}{x^3}$. If the particle is released from rest at a distance $2a$ from O show that it will come to rest instantaneously when $x = \frac{a}{\sqrt{2}}$, and find the time the particle takes to travel from $x = a$ to $x = a\sqrt{2}$.

$$\left[\textbf{Ans: } \frac{1}{2}a\sqrt{\left(\frac{a}{\mu}\right)}\right]$$

20. One end of an elastic string whose modulus of elasticity is λ, and natural length a, is tied to a fixed point on a smooth horizontal table, and the other end is tied to a mass m lying on the table. The particle is pulled to a distance where the extension of the string becomes b, and then is released; describe the character of the motion and show that the period of one complete oscillation is

$$2\left(\pi + \frac{a}{b}\right)\sqrt{\left(\frac{am}{\lambda}\right)}.$$

21. Show that the time of descent to the centre of force, the force varying inversely as the square of the distance from the centre, through the first half of its initial distance is to that through the second half as $\pi + 2 : \pi - 2$.

22. A particle whose mass is m is acted upon by a force $m\mu\left(x+\frac{a^4}{x^3}\right)$ towards the origin; if it starts from rest at a distance a, show that it will arrive at the origin in time $\frac{\pi}{4\sqrt{\mu}}$.

23. A particle falls from rest from a height b from the centre of the earth. Show that the velocity on reaching the centre is $\sqrt{\left[ga\left(3-\frac{2a}{b}\right)\right]}$, where g is gravity on the surface of the earth and a its radius.

24. If the velocity acquired by a particle on reaching the surface of the earth, when dropped from a height h_1, (supposing earth's attraction to be constant) is equal to the corresponding velocity acquired when dropped from a height h_2 (supposing the attraction to vary) then prove that

$$\frac{1}{h_1}-\frac{1}{h_2}=\frac{1}{a},$$

where a is earth's radius.

25. A particle falls towards the earth from infinity; show that its velocity on reaching the surface of the earth is same as that which it would have acquired in falling with constant acceleration g through a distance equal to the earth's radius.

26. A point moves in a straight line towards a centre of force $\frac{\mu}{(\text{Distance})^3}$, starting from the rest position with a distance a, from the centre of the force. Show that the time of reaching a point distance b from the centre of force is $a\sqrt{\left[\frac{(a^2-b^2)}{\mu}\right]}$, and that the velocity then is $\frac{\sqrt{\mu(a^2-b^2)}}{ab}$.

27. A particle moves in a straight line under a force to a point in it varying as $(\text{Distance})^{-4/3}$; show that the velocity in falling from rest at infinity to a distance a is equal to that acquired in falling from rest at a distance a to a distance $\frac{a}{8}$.

28. A particle of mass m is projected vertically under gravity, the resistance of the air being mk times the velocity. Show that the greatest height attained by the particle is $\left(\frac{V^2}{g}\right)[\lambda-\log(1+\lambda)]$, where V is the terminal velocity of the particle and λV its initial velocity.

29. A particle is projected vertically upwards with velocity u and the resistance of the air produces a retardation kv^2, where v is the velocity. Show that the velocity V with which the particle will return to the point of projection is given by

$$\frac{1}{V^2} = \frac{k}{g} + \frac{1}{u^2}.$$

30. If E is the kinetic energy of the particle in its upward path at a given point, show that the loss of energy when it passes the same point on the way down is $\frac{E^2}{(E+E')}$, where E' is the limit to which its energy approaches in its downward course.

31. A particle is allowed to fall from rest in a medium the resistance of which varies as the square of the velocity. When it has fallen through a distance a, a precisely similar particle is released from the same point as the first. Show that the ultimate distance between the particles is b, where $\frac{ag}{U^2} = \log\left[\cosh\left(\frac{bg}{U^2}\right)\right]$, U being the terminal velocity for either of the particles in the medium.

32. Prove that the time required by a body to reach a given fraction f of its terminal velocity, when it is dropped from rest and is acted on by gravity and a resistance proportional to any power of the velocity, is proportional to the terminal velocity.

33. A particle moves under gravity in a medium which offers a resistance $\frac{kv^2}{(a+y)}$ per unit mass, where v is the speed of the particle, y is the height above a fixed point O, and a and $k\left(\neq -\frac{1}{2}\right)$ are constants. If the particle is projected vertically upwards from O with speed u, show that it will come to rest instantaneously when $y = h$, where

$$(a+h)^{2k+1} = a^{2k+1}\left[1 + u^2\,\frac{(2k+1)}{2ag}\right].$$

Find also the speed V with which the particle will return to O.

$$\left[\textbf{Ans:}\quad V^2 = \frac{2ga}{2k-1}\left[1 - \left(1 + \frac{h}{a}\right)^{1-2k}\right]\right]$$

34. A particle is projected vertically upwards; the resistance of the air being assumed of the form mcv^2, where c is a constant and v is the velocity. During its motion, the particle has equal velocities at two points whose heights above the point of projection are x while moving upwards and y while moving downwards. Show that

$$e^{2c(a-x)} + e^{-2c(a-y)} = 2,$$

where a is the greatest height.

35. A rocket whose mass at time t is $m(1 - \alpha t)$, where m and α are constants, travels vertically upwards from rest at $t = 0$. The matter emitted has constant backward speed $\frac{4g}{\alpha}$ relative to the rocket. Assuming that the gravitational field g is constant and that the resistance of the atmosphere is $2mv\alpha$, where v is the speed of the rocket, show that half of the original mass is left when the rocket reaches a height $\left(\frac{g}{3\alpha^2}\right)$.

36. A rocket, initially of total mass M, throws off every second a mass fM with constant velocity V relative to the rocket. Show that it cannot rise at once unless $fV > g$ and that it cannot rise at all unless $fV > \lambda g$, where λM is the mass of the case of the rocket. If the conditions are such that the rocket is just able to rise vertically at once, show that the greatest height it will reach is

$$\frac{V^2}{g}\left[1 - \lambda + \log\lambda + \frac{1}{2}(\log\lambda)^2\right].$$

37. From a rocket which is free to move vertically upwards, matter is ejected downwards with constant relative velocity gT at a constant rate $\frac{2M}{T}$. Initially the rocket is at rest and has mass $2M$, half of which is available for ejection. Neglecting air resistance and variations in gravitational attraction, show that the greatest upward speed is attained when the mass of the rocket is reduced to M, and determine this speed. Show that the rocket rises to a height $\frac{1}{2}gT^2(1 - \log 2)^2$.

$$\left[\textbf{Ans:}\quad \text{Max. speed} = gT\left(\log 2 - \frac{1}{2}\right)\right]$$

38. A rocket is launched in a vertical direction. Its initial mass in M_0 and its mass at burnout is $\frac{1}{4}M_0$. The engines burn for 100 s and the constant speed at which the gas is expelled relative to the rocket is 3000 m/sec. Calculate rocket's speed at burnout.

If the mass of the rocket decreases at a constant rate with respect to time during the burn, determine the altitude of the rocket at burnout and find the constant value of the thrust of the engine during the burn.

[**Ans:** 3180 m/sec; 112 km; 2.4 $M_0 g$]

39. A rocket is launched in a vertical direction. Its initial mass is M_0 and its mass at burnout is $\left(\frac{1}{2}\right)M_0$. The engines burn for 42 sec and the constant speed at which the gas is expelled relative to the rocket is 2500 m/sec. Calculate the rocket's speed at burnout.

[**Ans:** 1320 m/sec]

40. A lunar module of total mass M is at a height H above the surface of the moon and is descending vertically with speed V, when a rocket is ignited to produce a soft landing. The mass of the fuel decreases at a constant rate, and the gas is ejected at a speed of 2400 m/sec relative to the module. If the module touches the lunar surface with zero velocity and the module's mass at the end of the burn lasting 350 s is $\left(\frac{2}{3}\right)M$, evaluate V and H. (Assume that the acceleration due to gravity at the lunar surface is 1.62 m/s^2).

[**Ans:** 306 m/s; 82.6 km]

Kinematics in Two Dimensions

7.1 INTRODUCTION

There are many problems of interest when motion is confined to a plane, e.g., motion of a projectile, particle moving in contact with a plane curve and motion of planets and artificial satellites.

Displacement, velocity and acceleration are basic quantities in the study of a motion of a particle. These are linked to the position of the particle along its path. Trajectory or path is a geometrical quantity and determination of the link, through mathematical formulas expressing velocity and acceleration components in various co-ordinate systems, is the concern of kinematics.

7.2 CARTESIAN CO-ORDINATES

The simplest co-ordinate system to write down these relations is the rectangular cartesian system but it must be borne in mind that it may not be the best suited to deal with the problem in hand. In Cartesian co-ordinates, the position of a particle in the three-dimensional space is specified by the position vector.

$$\mathbf{r} = x\hat{\mathbf{i}} + y\hat{\mathbf{j}} + z\hat{\mathbf{k}}, \tag{7.1}$$

where $\hat{\mathbf{i}}, \hat{\mathbf{j}}$ and $\hat{\mathbf{k}}$ are unit vectors along x, y and z directions, respectively. Thus, we can easily write

$$\mathbf{v} = \frac{d\mathbf{r}}{dt} = \frac{dx}{dt}\hat{\mathbf{i}} + \frac{dy}{dt}\hat{\mathbf{j}} + \frac{dz}{dt}\hat{\mathbf{k}} \tag{7.2}$$

$$\mathbf{f} = \frac{d\mathbf{v}}{dt} = \frac{d^2x}{dt^2}\hat{\mathbf{i}} + \frac{d^2y}{dt^2}\hat{\mathbf{j}} + \frac{d^2z}{dt^2}\hat{\mathbf{k}}. \tag{7.3}$$

The velocity components (v_x, v_y, v_z) and acceleration components (f_x, f_y, f_z) are readily seen to be

$$v_x = \frac{dx}{dt}, v_y = \frac{dy}{dt}, v_z = \frac{dz}{dt} \tag{7.4}$$

$$f_x = \frac{d^2x}{dt^2}, f_y = \frac{d^2y}{dt^2}, f_z = \frac{d^2z}{dt^2}. \tag{7.5}$$

It may also be seen that we can write the acceleration components as

$$f_x = \frac{dv_x}{dt} = v_x\frac{dv_x}{dx} = \frac{1}{2}\frac{dv_x^2}{dx}. \tag{7.6}$$

In the case of uniplanar motion in the x, y plane, the quantities are independent of z and there is no component of a vector in z-direction. Thus, we have

$$\mathbf{r} = x\hat{\mathbf{i}} + y\hat{\mathbf{j}}, \tag{7.7}$$

$$\mathbf{r} = \frac{d\mathbf{r}}{dt} = \frac{dx}{dt}\hat{\mathbf{i}} + \frac{dy}{dt}\hat{\mathbf{j}}, \tag{7.8}$$

$$\mathbf{f} = \frac{d^2x}{dt^2}\hat{\mathbf{i}} + \frac{d^2y}{dt^2}\hat{\mathbf{j}}. \tag{7.9}$$

The discerning reader would have observed that starting from the position vector **r**, the components of velocity and acceleration were easily derived, because in Cartesian co-ordinates the base vectors $\hat{\mathbf{i}}, \hat{\mathbf{j}}$ and $\hat{\mathbf{k}}$ are independent of the position of the field point. But this will not be so in other co-ordinate systems, as we shall see in the following two sections where we have confined ourselves to uniplanar motion.

7.3 INTRINSIC CO-ORDINATES

The natural co-ordinate system for studying the motion of a particle along a curve is the intrinsic co-ordinate system (s, ψ) where s is the distance measured along the arc from a fixed point A to the variable point P, and ψ is the slope of the tangent there. The two unit base vectors in this system are $\hat{\mathbf{t}}$ (the unit tangent vector along the direction of increasing s) and $\hat{\mathbf{n}}$, the unit normal vector along the principal normal which may be noted to follow the inward direction (refer Figure 7.1).

Now, from the definition of velocity, with P and the neighbouring point Q on the path being represented **r** by and $\mathbf{r} + \delta\mathbf{r}$, respectively, and δs the arc distance PQ, we have

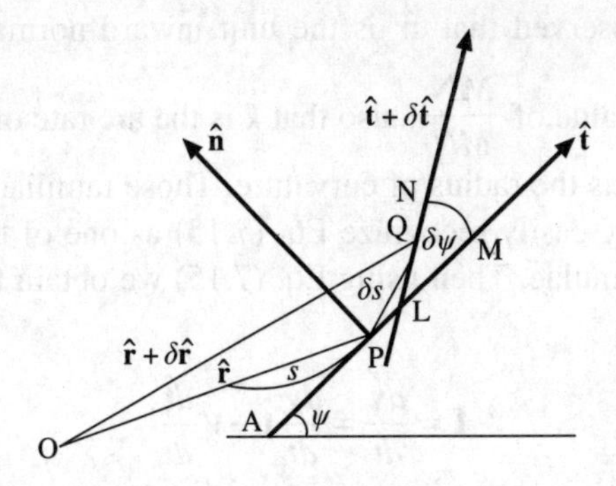

Figure 7.1 Geometry of the particle P moving on the path PQ.

$$\mathbf{v}=\frac{d\mathbf{r}}{dt}=\lim_{\delta t\to 0}\frac{\delta\mathbf{r}}{\delta t} \tag{7.10}$$

$$=\lim_{Q\to P}\frac{\text{chord } PQ}{\text{arc } PQ}\times\lim\frac{\mathbf{PQ}}{\text{PQ}}\times\frac{\delta s}{\delta t}$$

$$=1\times\hat{\mathbf{t}}\times\frac{ds}{dt}=\frac{ds}{dt}\hat{\mathbf{t}}$$

$$=v\hat{\mathbf{t}},$$

where

$$v=|\,\mathbf{v}\,|=\frac{ds}{dt}=\dot{s}\ \text{ is the speed of the particle.} \tag{7.11}$$

In deriving Eq. (7.11), we have made use of the fact that

$$\lim_{Q\to P}\frac{\text{chord } PQ}{\text{arc } PQ}=1 \tag{7.12}$$

and that in the limit the chord assumes the position of the tangent. The above result (Eq. 7.10) shows that the motion is wholly along the curve. Thus, we have

$$\text{Tangential velocity}:\ v_t=v=|\mathbf{v}|, \tag{7.13}$$

$$\text{Normal velocity}:\ v_n=0. \tag{7.14}$$

Since acceleration involves the rate of change of velocity, we shall first write down the arc rate of change of unit tangent vector $\hat{\mathbf{t}}$. Referring to Figure 7.1, with arc MN as an arc of a circle of unit radius, and $\delta\psi$ as the angle through which the tangent at P turns while going from P to Q, we have

$$\frac{d\hat{\mathbf{t}}}{ds}=\lim_{\delta s\to 0}\frac{\delta\hat{\mathbf{t}}}{\delta s}=\lim_{N\to M}\frac{\mathbf{MN}}{\delta s} \tag{7.15}$$

$$=\lim_{N\to M}\frac{\text{chord } MN}{\text{arc } MN}\lim_{\delta s\to 0}\frac{1\delta\psi}{\delta s}\lim_{N\to M}\frac{\mathbf{MN}}{\text{MN}}$$

$$=\frac{\delta\psi}{\delta s}\hat{\mathbf{n}}=k\hat{\mathbf{n}}=\frac{1}{\rho}\hat{\mathbf{n}}.$$

Here, it is to be observed that $\hat{\mathbf{n}}$ is the unit inward normal to the path at P, being the limiting value of $\frac{\mathbf{MN}}{MN}$; also that k is the arc rate of turning of tangent and its reciprocal ρ is the radius of curvature. Those familiar with the geometry of space curves may easily recognize Eq. (7.15) as one of the results included in Frenet-Serret formulae. Then using Eq. (7.15) we obtain from Eq. (7.10) the acceleration

$$\mathbf{f} = \frac{d\mathbf{v}}{dt} = \frac{dv}{dt}\hat{\mathbf{t}} + v\frac{d\hat{\mathbf{t}}}{dt} \tag{7.16}$$

$$= \frac{dv}{dt}\hat{\mathbf{t}} + v\frac{ds}{dt}\frac{d\hat{\mathbf{t}}}{ds}$$

$$= \frac{dv}{dt}\hat{\mathbf{t}} + \frac{v^2}{\rho}\hat{\mathbf{n}}.$$

Thus, we have
tangential acceleration:

$$f_t = \frac{dv}{dt} = \frac{dv}{dt}\frac{ds}{dt} = v\frac{dv}{ds} = \frac{1}{2}\frac{dv^2}{ds} = \frac{d^2s}{dt^2} = \ddot{s}, \tag{7.17}$$

normal acceleration:

$$f_n = \frac{v^2}{\rho} = kv^2 = \frac{ds}{dt}\frac{d\psi}{dt} = \rho\left(\frac{d\psi}{dt}\right)^2. \tag{7.18}$$

Equations (7.16) – (7.18) show that the acceleration of the particle always lies in the osculating plane, the tangential acceleration is independent of the shape of the curve and normal acceleration depends on the curvature. It may be noted that the Eqs. (7.10) and (7.16) are true for the motion of a particle along a space curve which need not be a plane curve.

In the case of uniform motion in a circle $s = a\psi$, the tangential acceleration vanishes and we always have the constant normal acceleration $\frac{v^2}{a}$ pointing towards the centre which is called *centripetal acceleration*, a name often used for $\frac{v^2}{\rho}$ in the general case. (The corresponding force $-\frac{v^2}{\rho}$ directed away from the centre is the centrifugal force).

7.4 POLAR CO-ORDINATES

The position of a particle moving in a plane may be specified by

$$\mathbf{r} = r\hat{\mathbf{r}} \tag{7.19}$$

where r and θ are the polar co-ordinates of the moving point, and $\hat{\mathbf{r}}$ and $\hat{\boldsymbol{\theta}}$, the unit vectors along the radial and the transverse directions respectively. In contrast to $\hat{\mathbf{i}}$ and $\hat{\mathbf{j}}$ the vectors $\hat{\mathbf{r}}$ and $\hat{\boldsymbol{\theta}}$ are not fixed in directions. We first calculate their rates of change.

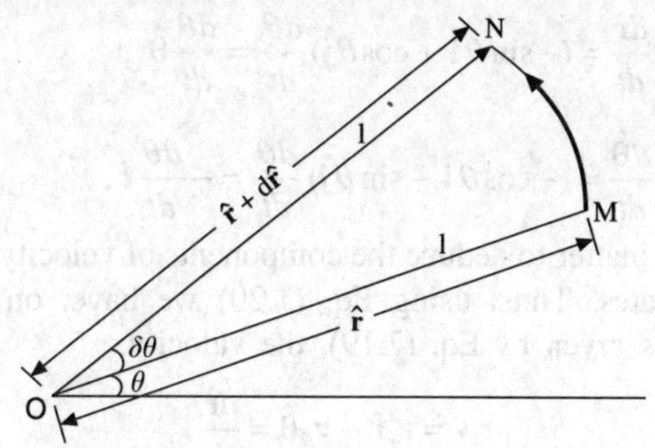

Figure 7.2 Rate of change of unit vector $\hat{\mathbf{r}}$.

Since $\mathbf{r}$ is a vector of constant magnitude, the vector $\frac{d\mathbf{r}}{dt}$ is perpendicular o it. In Figure 7.2 let OM be unit vector $\hat{\mathbf{r}}$, ON the unit vector $\hat{\mathbf{r}} + \delta\hat{\mathbf{r}}$, $\delta\theta$ the nfinitesimal angle included between OM and ON. Then arc MN = $\delta\theta$, being rc of a unit circle. Hence, we have

$$\frac{d\hat{\mathbf{r}}}{dt} = \lim_{\delta t\to 0} \frac{\delta\hat{\mathbf{r}}}{\delta t} = \lim_{\delta t\to 0} \frac{\mathbf{MN}}{\delta t} \tag{7.20}$$

$$= \lim_{N\to M} \frac{\text{chord } MN}{\text{arc } MN} \lim_{\delta t\to 0} \frac{\delta\theta}{\delta t} \frac{\mathbf{MN}}{\text{MN}} \tag{7.20a}$$

$$= \frac{d\theta}{dt}\hat{\boldsymbol{\theta}}. \tag{7.20b}$$

n Eq. (7.20a), the unit vector is obtained as chord MN ultimately becomes erpendicular to OM as $\delta\theta \to 0$. The students are advised to derive, in a similar vay, the following result

$$\frac{d\hat{\boldsymbol{\theta}}}{dt} = -\frac{d\theta}{dt}\hat{\mathbf{r}}. \tag{7.21}$$

Vriting $\frac{d\theta}{dt} = \omega$ for the angular velocity, Eqs. (7.20), (7.20a) and (7.20b) are lso used in the form

$$\frac{d\hat{\mathbf{r}}}{dt} = \omega\hat{\boldsymbol{\theta}}, \frac{d\hat{\boldsymbol{\theta}}}{dt} = -\omega\hat{\mathbf{r}}. \tag{7.22}$$

emark: We can obtain the result obtained in Eqs. (7.20) and (7.21) by riting down the unit vectors $\hat{\mathbf{r}}$ and $\hat{\boldsymbol{\theta}}$ in terms of the unit vectors $\hat{\mathbf{i}}$ and $\hat{\mathbf{j}}$, nd then differentiating.

$$\hat{\mathbf{r}} = \cos\theta\,\hat{\mathbf{i}} + \sin\theta\,\hat{\mathbf{j}},\ \hat{\boldsymbol{\theta}} = -\sin\theta\,\hat{\mathbf{i}} + \cos\theta\,\hat{\mathbf{j}},$$

and hence,

$$\frac{d\hat{\mathbf{r}}}{dt} = (-\sin\theta\,\hat{\mathbf{i}} + \cos\theta\,\hat{\mathbf{j}})\frac{d\theta}{dt} = \frac{d\theta}{dt}\hat{\boldsymbol{\theta}}$$

$$\frac{d\hat{\boldsymbol{\theta}}}{dt} = (-\cos\theta\,\hat{\mathbf{i}} - \sin\theta\,\hat{\mathbf{j}})\frac{d\theta}{dt} = -\frac{d\theta}{dt}\hat{\mathbf{r}}.$$

Now, it is an easy matter to deduce the components of velocity and acceleration in polar co-ordinates. Thus, using Eq. (7.20) we have, on taking the time derivative of $\hat{\mathbf{r}}$ as given by Eq. (7.19), the velocity.

$$\mathbf{v} = v_r\hat{\mathbf{r}} + v_\theta\hat{\boldsymbol{\theta}} = \frac{d\hat{\mathbf{r}}}{dt} \tag{7.23}$$

$$= \frac{dr}{dt}\hat{\mathbf{r}} + r\frac{d\theta}{dt}\hat{\boldsymbol{\theta}},$$

providing

$$\text{Radial velocity}: \ v_r = \frac{dr}{dt} = \dot{r} \tag{7.24}$$

$$\text{Transverse velocity}: \ v_\theta = r\frac{d\theta}{dt} = r\omega = r\dot{\theta}. \tag{7.25}$$

Further differentiation and use of Eqs. (7.20) and (7.21) yields acceleration

$$\mathbf{f} = f_r\hat{\mathbf{r}} + f_\theta\hat{\boldsymbol{\theta}} = \frac{d\mathbf{v}}{dt} = \frac{d}{dt}(v_r\hat{\mathbf{r}} + v_\theta\hat{\boldsymbol{\theta}}) \tag{7.26}$$

$$= \frac{d^2r}{dt^2}\hat{\mathbf{r}} + \frac{dr}{dt}\frac{d\hat{\mathbf{r}}}{dt} + \frac{d}{dt}\left(r\frac{d\theta}{dt}\right)\hat{\boldsymbol{\theta}} + r\frac{d\theta}{dt}\frac{d\hat{\boldsymbol{\theta}}}{dt}$$

$$= \left[\frac{d^2r}{dt^2} - r\left(\frac{d\theta}{dt}\right)^2\right]\hat{\mathbf{r}} + \left[\frac{dr}{dt}\frac{d\theta}{dt} + \frac{d}{dt}\left(r\frac{d\theta}{dt}\right)\right]\hat{\boldsymbol{\theta}}.$$

The radial acceleration is given by

$$f_r = \frac{d^2r}{dt^2} - r\left(\frac{d\theta}{dt}\right)^2 = \frac{d^2r}{dt^2} - (r\omega^2) = \ddot{r} - r\dot{\theta}^2. \tag{7.27}$$

Transverse acceleration is given by

$$f_\theta = \frac{1}{r}\frac{d}{dt}\left(r^2\frac{d\theta}{dt}\right) = \frac{1}{r}\frac{d}{dt}(r^2\omega) = r\ddot{\theta} + 2\dot{r}\dot{\theta}. \tag{7.28}$$

For uniform motion in a circle $r = a$ = Constant and ω = Constant consequently then we have

$$v_r = 0, v_\theta = a\omega, f_r = -a\omega^2, f_\theta = 0. \tag{7.29}$$

Thus, once again we see that the particle moves with an inward centripetal acceleration $a\omega^2 = \dfrac{v^2}{a}$, where $v = v_\theta = a\omega$ is the speed in this case.

Uniform motion in a circle may occur under the influence of a constant external force T in the radial direction. (The student should ponder why the motion is not rectilinear). The equation of motion is

$$-r\omega^2 = T \tag{7.30}$$

For example for a particle of unit mass tied to one end of a string, the other end of it being fixed at the centre of the circle on which the particle moves. T is the force of tension acting towards the centre. On the system, i.e., the particle, it acts as external force, although actually it is an internal force. Equation (7.30) may be expressed as

$$r\omega^2 + T = 0 \tag{7.31}$$

and then may be looked upon as a static equation of equilibrium (in a co-ordinate system rotating with angular velocity ω) with $r\omega^2$ interpreted as centrifugal force acting in the outward direction. But it should be kept in mind that centrifugal force is only a fictitious force.

SOLVED EXAMPLES

EXAMPLE 7.1 A point moves in a curve so that its tangential and normal accelerations are equal and the angular velocity of the tangent is constant. Find the path.

Solution We are given

$$\frac{d^2s}{dt^2} = \frac{v^2}{\rho} = \left(\frac{ds}{dt}\right)^2 \frac{d\psi}{ds} \tag{i}$$

and

$$\frac{d\psi}{dt} = \alpha = \text{Constant}. \tag{ii}$$

From Eq. (ii) we obtain

$$\frac{ds}{dt} = \frac{ds}{d\psi}\frac{d\psi}{dt} = \alpha\frac{ds}{d\psi} \tag{iii}$$

and

$$\frac{d^2s}{dt^2} = \alpha^2\frac{d^2s}{d\psi^2}. \tag{iv}$$

Substituting from Eqs. (iii) and (iv) into Eq. (i), we get

$$\frac{d^2s}{d\psi^2} = \frac{ds}{d\psi}. \tag{v}$$

Integrating Eq. (v) we get the path

$$s = Ae^{\psi} + B.$$

EXAMPLE 7.2 A particle is moving in a parabola with uniform angular velocity about the focus; prove that its normal acceleration at any point is proportional to the radius of curvature of its path at that point.

Solution We are given

$$\frac{d\theta}{dt} = \omega = \text{Constant}. \tag{i}$$

Also the path, given to be a parabola, may be represented in polar co-ordinates as

$$\frac{L}{r} = 1 + \cos\theta. \tag{ii}$$

This provides

$$\tan\phi = r\frac{d\theta}{dr} = \frac{1+\cos\theta}{\sin\theta} = \cot\frac{\theta}{2}$$

or
$$\phi = \frac{\pi}{2} - \frac{\theta}{2}.$$

Therefore,
$$\psi = \theta + \phi = \frac{\pi}{2} + \frac{\theta}{2}. \tag{iii}$$

Now, we have

$$\frac{v^2}{\rho} = \frac{1}{\rho}\left(\frac{ds}{dt}\right)^2 = \frac{1}{\rho}\left(\frac{ds}{d\psi}\right)^2\left(\frac{d\psi}{dt}\right)^2$$

$$= \frac{1}{4}\rho \qquad \text{[From Eq. (ii)]}$$

showing that normal acceleration is proportional to the radius of curvature.

EXAMPLE 7.3 If the tangential and normal accelerations of a particle describing a plane curve is constant throughout the motion, prove that the angle ψ through which the direction of motion turns in time t is given by $\psi = k \log (1 + At)$.

Solution We are given

Tangential acceleration:
$$\frac{d^2 s}{dt^2} = \alpha = \text{Constant}, \tag{i}$$

Normal acceleration:
$$\frac{ds}{dt}\frac{d\psi}{dt} = \beta = \text{Constant}, \tag{ii}$$

First integral of Eq. (i) is

$$\frac{ds}{dt} = \alpha t + \gamma. \tag{iii}$$

Equations (ii) and (iii) now provide

$$d\psi = \frac{\beta\, dt}{\alpha t + \gamma} \tag{iv}$$

$$\psi = \frac{\beta}{\alpha} \log\left(1 + \frac{\alpha}{\gamma} t\right). \tag{v}$$

where we have taken $\psi = 0$ at $t = 0$. Renaming the constants we get the answer in the required form.

EXAMPLE 7.4 The velocities of a particle along and perpendicular to the radius from a fixed origin are λr and $\lambda a\theta$; find the path, and show that the accelerations, along and perpendicular to the radius vector, are

$$\lambda^2\left(r - \frac{a^2}{r}\theta^2\right) \text{ and } \lambda^2 a\theta\left(1 + \frac{a}{r}\right).$$

Solution We are given

$$\frac{dr}{dt} = \lambda r \tag{i}$$

$$r\frac{d\theta}{dt} = \lambda a\theta. \tag{ii}$$

From Eqs. (i) and (ii), we have

$$a\frac{dr}{r^2} = \frac{d\theta}{\theta},$$

which on integration yields the path as

$$\frac{a}{r} = K - \log\theta. \tag{iii}$$

Again, on making use of Eqs. (i) and (ii), we have
Radial acceleration:

$$f_r = \frac{d^2 r}{dt^2} - r\left(\frac{d\theta}{dt}\right)^2 = \frac{d}{dt}(\lambda r) - r\frac{\lambda^2 a^2}{r^2}$$

$$= \lambda^2\left(r - \frac{a^2\theta^2}{r}\right).$$

Transverse acceleration:

$$f_\theta = \frac{1}{r}\frac{d}{dt}\left(r^2\frac{d\theta}{dt}\right) = \frac{1}{r}\frac{d}{dt}(\lambda a r\theta)$$

$$= \frac{\lambda a}{r}\left(\frac{dr}{dt}\theta + r\frac{d\theta}{dt}\right) = \lambda^2 a\left(\theta + \frac{a}{r}\right).$$

EXAMPLE 7.5 If the angular velocity ω of a particle about the origin is constant, and the rate of change of acceleration is directed wholly along the radius vector, prove that $\frac{d^2r}{dt^2} = \frac{1}{3}\omega^2 r.$

Solution We know that

Angular velocity:
$$\frac{d\theta}{dt} = \omega = \text{Constant}. \qquad \text{(i)}$$

Now, the rate of change of acceleration is

$$\frac{d\mathbf{f}}{dt} = \left(\frac{df_r}{dt} - \omega f_\theta\right)\hat{\mathbf{r}} + \left(\frac{df_\theta}{dt} - \omega f_r\right)\hat{\boldsymbol{\theta}}, \qquad \text{(ii)}$$

where use has been made of Eqs. (7.20) and (7.21).

Since this is given to be directed wholly along the radius vector, $\hat{\boldsymbol{\theta}}$ component vanishes, and hence

$$\frac{df_\theta}{dt} + \omega f_r = 0. \qquad \text{(iii)}$$

Next, we make use of Eqs. (7.27), (7.28) and Eq. (i) to express Eq. (iii) as

$$\frac{d}{dt}\left(2\omega\frac{dr}{dt}\right) + \omega\left(\frac{d^2r}{dt^2} - r\omega^2\right) = 0$$

or
$$\frac{d^2r}{dt^2} = \frac{1}{3}\omega^2 r.$$

EXAMPLE 7.6 A particle describes a cardioid $r = a(1 + \cos\theta)$ in such a manner that the radius vector from the origin rotates with uniform angular velocity ω. Show that the acceleration consists of a component $2a\omega^2$ parallel to the initial line and a component $(4r - 3a)\omega^2$ towards the origin.

Solution The path is

$$r = a(1 + \cos\theta). \qquad \text{(i)}$$

Therefore, we have

$$\dot{r} = -a\omega\sin\theta, \qquad \text{(ii)}$$

$$\ddot{r} = -a\omega^2\cos\theta. \qquad \text{(iii)}$$

Now, from Eqs. (7.27) and (7.28), we write

$$f_r = \ddot{r} - a\omega^2 = -a\omega^2(1 + 2\cos\theta), \qquad \text{(iv)}$$

$$f_\theta = 2\dot{r}\omega + r\dot{\omega} = -2a\omega^2\sin\theta. \qquad \text{(v)}$$

If f_1 and f_0 represent the components of the acceleration parallel to the initial line and the component towards the origin respectively, we have

$$f_1 \cos\theta - f_o = f_r \quad \text{(vi)}$$

$$f_1 \sin\theta = -f_\theta \quad \text{(vii)}$$

$$f_1 = 2a\omega^2. \quad \text{(viii)}$$

and then from Eqs. (iv), (vi), (viii) and (i), we obtain

$$f_o = a\omega^2(1 + 3\cos\theta) = (4r - 3a)\,\omega^2.$$

PROBLEMS

1. A particle describes the equiangular spiral $r = ae^\theta$ in such a manner that its acceleration has no radial component. Prove that the magnitudes of the velocity and acceleration are each proportional to r.
2. A curve is described by a particle having a constant acceleration in a direction inclined at a constant angle to the tangent; show that the curve is an equiangular spiral.
3. A point moves along the arc of a cycloid in such a manner that the tangent at it rotates with constant angular velocity. Show that the acceleration of the moving point is constant in magnitude.
4. A point moves in a plane curve and at time t passes through the point of the curve at which curvature is k with velocity v. If $\dfrac{d^2v}{dt^2} = av$, where a is a constant, show that $k = \dfrac{b}{v^3}$, where b is a constant, and that $\dot{v}^2 = av^2 + c - \dfrac{b^2}{v^2}$, where c is a constant.
5. A point moves in a parabola with constant speed. Show that the angular velocity about the focus varies as $\cos^3\left(\dfrac{\theta}{2}\right)$, whose θ is the angular distance from the vertex.
6. A point moves in a curve with constant tangential acceleration and the magnitudes of the tangential velocity and the normal accelerations are in constant ratio. Find the intrinsic equation of the curve.

   [**Ans:** $s = k(\psi + \alpha)^2 + C$]
7. Prove that the angular acceleration of the direction of motion of a moving point in a plane is

$$\frac{v}{\rho}\frac{dv}{ds} - \frac{v^2}{\rho^2}\frac{d\rho}{ds}$$

8. A particle describes a curve with uniform speed v. If the acceleration at any point s be $\dfrac{cv^2}{(s^2+c^2)}$; prove that the curve is a catenary.

9. A particle, projected with a velocity u, is acted upon by a force which produces a constant acceleration f in the plane of the motion inclined at a constant angle α with the direction of the motion. Obtain the intrinsic equation of the curve described and show that the particle will be moving in the opposite direction to that of projection at time $\dfrac{u}{f\cos\alpha}(e^{\pi\cot\alpha}-1)$.

10. A particle moves in the curve $y = a\log\left(\sec\dfrac{x}{a}\right)$ in such a way that the tangent to the curve rotates uniformly, prove that the resultant acceleration of the particle varies as the square of the radius of curvature.

11. A particle moves in a catenary ($s = c\tan\psi$), and the direction of its acceleration at any point makes equal angles with the tangent and the normal to the path at that point. If the speed at the vertex be u, show that the velocity and acceleration at any other point are given by $u = e^{\psi}$ and $\dfrac{\sqrt{2}}{c}u^2e^{2\psi}\cos^2\psi$.

12. A particle moves with constant speed v along the cardioid $r = a(1 + \cos\theta)$. Show that the radial component of the acceleration is constant, and that both $\dfrac{d\theta}{dt}$ and the magnitude of the resultant acceleration are proportional to $r^{\frac{-1}{2}}$.

13. The velocities of a particle along and perpendicular to a radius vector from a fixed origin are λr^2 and $\mu\theta^2$, where λ and μ are constants; find the polar equation of the path of the particle and also its radial and transverse accelerations in terms of r and θ.

$$\left[\textbf{Ans: } \frac{1}{r^2} = \frac{2\lambda}{\mu\theta} + C;\ 2\lambda^2r^2 - \frac{\mu^2\theta^4}{r},\ \lambda\mu r\theta^2 + \frac{2\mu^2\theta^2}{r}\right]$$

14. A point describes the curve whose polar equation is $r^2 = a + b\cos 2\theta$, $a > b$, so that the time t is given by $2ht = 2a\theta + b\sin 2\theta$, where a, b and h are constants. Prove that the acceleration is radial and express it as a function of r only.

$$\left[\textbf{Ans: } \frac{3h^2(a^2-b^2)}{r^7} - \frac{4ah^2}{r^5}\right]$$

15. If a rod which always passes through the origin rotates with the uniform angular velocity ω, while one end describes the curve $r = a + be^{\theta}$; show that the radial acceleration of any point of the rod is the same at every instant, and the radial velocity is the same at every point at a given instant.

16. A smooth straight wire rotates in a horizontal plane with constant angular velocity about one end. Show that a particle which is free to slip along the wire may describe an equiangular spiral.

17. Show that the path of a point P which possesses two constant velocities u and v, the first of which is in a fixed direction and the other is perpendicular to the radius OP drawn from a fixed point O, is a conic whose focus is O and whose eccentricity is $\dfrac{u}{v}$.

18. A vessel steams at a constant speed u along a straight line whilst another vessel, steaming at a constant speed v, always moves at right angles to the radius vector to the former. Show that the path of either vessel relative to the other is a conic section of eccentricity $\dfrac{u}{v}$.

19. A point describes a circle of diameter d with uniform speed u; show that the radial and transverse accelerations are

$$-\frac{2u^2}{d}\cos\theta, \quad -\frac{2u^2}{d}\sin\theta,$$

if a diameter is taken as the initial line and one end of the diameter as pole.

20. An aircraft pursues a straight course with velocity u and is being chased by a guided missile moving with constant speed $2u$ and fitted with a homing device to ensure that its motion is always directed at the target. Initially the missile is at right angles to the course of the aircraft and distant R from it. Find the polar equation of the missile's pursuit curve relative to the target, taking the course of the target as the initial line $\theta = 0$, and the time taken by it to strike the target.

$$\left[\textbf{Ans:}\quad \frac{R}{r} = 2\sin^3\frac{\theta}{2}\cos\frac{\theta}{2};\ \frac{2R}{3u}\right]$$

21. If the acceleration of a particle describing a curve in a plane is resolved into two components, one parallel to the initial line and the other along the radius vector. Prove that these components are

$$-\frac{1}{r\sin\theta}\frac{d}{dt}(r^2\omega) \text{ and } \frac{d^2r}{dt^2} - r\omega^2 + \frac{\cos\theta}{r}\frac{d}{dt}(r^2\omega),$$

where $$\omega = \frac{d\theta}{dt}.$$

22. A particle is moving with constant angular velocity $a\sqrt{3}$ about the origin in a plane. If the rate of change of acceleration is directed wholly along the radius vector, prove that

$$r = a\cosh \alpha t + b\sinh \alpha t,$$

where, initially the particle is at a distance a and moving with velocity αb along the radius vector.

23. A smooth straight tube OA is made to rotate with uniform angular velocity ω about a vertical axis OZ so that AOZ is a constant acute angle α. Prove that the particle in the tube will remain at rest relative to the tube at a distance $\left(\dfrac{g\cos\alpha}{\omega^2 \sin^2 \alpha}\right)$ from O.

24. A ring, which can slide on a thin long smooth rod, rests at a distance d from one end O. The rod is then set revolving uniformly about O in a horizontal plane, show that in space the ring describes the curve $r = d \cosh\theta$.

25. A point starts from the origin in the direction of the initial line with velocity $\dfrac{f}{\omega}$ and moves with constant angular velocity ω about origin and with constant negative radial acceleration $-f$. Show that the rate of growth of the radial velocity is never positive, but tends to the limit zero, and prove that the equation of the path is $\omega^2 r = f(1 - e^{-\theta})$.

26. (i) The earth moves around the sun in an almost circular orbit of radius 1.50×10^{11} m, in one year. Determine the speed and acceleration of the earth relative to a co-ordinate system with the sun as the origin.

(ii) Consider the earth as a sphere of radius 6.371×10^6 m and let its period of rotation about the polar axis be 86164 s. Determine the speed and acceleration at latitude λ about the polar axis.

[**Ans:** (i) 30×10^4 ms^{-1}, 0.0059 ms^{-2} towards the Sun, (ii) 465 cos λ ms^{-1}, 0.0339 cos λ ms^{-2}, towards the centre of the earth]

Constrained Motion

8.1 INTRODUCTION

Newton's laws of motion apply to a free particle. But the motion of a particle may be constrained by some imposed conditions restricting its movement, e.g., sliding down an elliptic wire in a vertical plane it is forced to take an elliptic path. The external body force, viz., gravity cannot account for this curved path. Here the particle is constrained to remain in contact with the wire; hence, invoking Newton's third law, the force of reaction comes into play and accounts for the elliptic path. It may be noted that this force is not prescribed, but emerges in the solution of the problem and hence may be termed a dynamic constraint. The imposed conditions, thus, are seen to manifest themselves in the appearance of forces called constraint reactions. In other words, the conditions of constraint are replaced by constraint reactions. Thus, a constrained particle may be treated as a free particle (i.e. without constraints) provided the constraint reactions are also included as body forces. It may be noted that although the motion here takes place in two dimensions, but because of the constraints, it is only with one degree of freedom.

8.2 MOTION ON A SMOOTH CURVE

Let a particle of mass m be constrained to move on a fixed smooth curve under the action of an external force $\mathbf{F}(x, y)$. The force of constraint is normal reaction R (which in the situation depicted in Figure 8.1 acts along the inward normal).

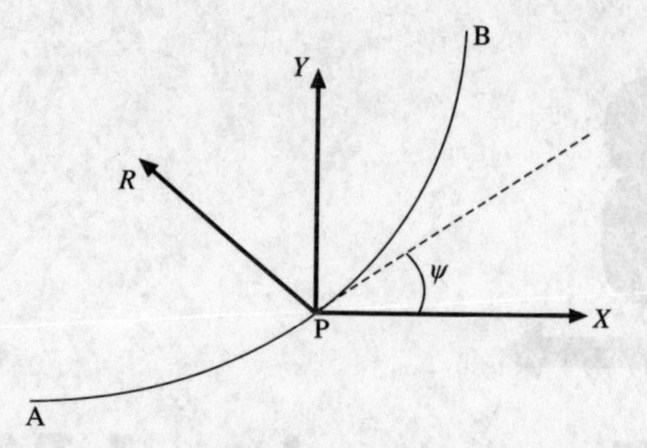

Figure 8.1 Constrained motion of a particle P on a smooth curve AB.

Using the expressions for tangential and normal components of acceleration the equations of motion may be written as

$$mv\frac{dv}{ds} = X\cos\psi + Y\sin\psi = F_S, \tag{8.1}$$

$$m\frac{v^2}{\rho} = R - X\sin\psi + Y\cos\psi = R + F_n. \tag{8.2}$$

where ψ is the slope of tangent at P.
Integrating Eq. (8.1), we obtain speed v of the particle as given by the equation

$$\frac{1}{2}m\left(v^2 - {v_A}^2\right) = \int_{S_A}^{S} F_S\, ds, \tag{8.3}$$

where A is some fixed point on the curve and v_A the speed there. Equation (8.3) may be written directly from the work-energy principle (refer Chapter 1).

Knowing v, we can get the unknown constraint reaction R from Eq.(8.2) as

$$R = \frac{mv^2}{\rho} + X\sin\psi - Y\cos\psi. \tag{8.4}$$

The particle will leave the curve when $R = 0$, and hence, from Eq. (8.4), it follows that then

$$\frac{mv^2}{\rho} = Y\cos\psi - X\sin\psi = F_n. \tag{8.5}$$

If the only external force is the force of gravity then $X = 0$ and $Y = -mg$. Equation (8.3) in this case provides

$$v^2 = v_A^2 - 2g\,(y - y_A) = v_A^2 - 2gh, \tag{8.6}$$

where h is the vertical distance moved by the particle; Eq. (8.4) then becomes

$$\frac{R}{m} = \frac{v^2}{\rho} + g\cos\psi, \tag{8.7}$$

It follows that the particle leaves the curve when

$$v^2 = -\rho g \cos\psi = -g\frac{dx}{d\psi}. \tag{8.8}$$

8.3 CIRCULAR MOTION IN A VERTICAL PLANE

We shall consider below three types of circular motion exhibited by following three problems:

Problem 1

A particle is projected inside a fixed smooth cylinder or sphere, with circular section in a vertical plane, from the lowest point A with initial horizontal velocity u. To discuss the motion.

Referring Figure 8.2, we have from energy equation

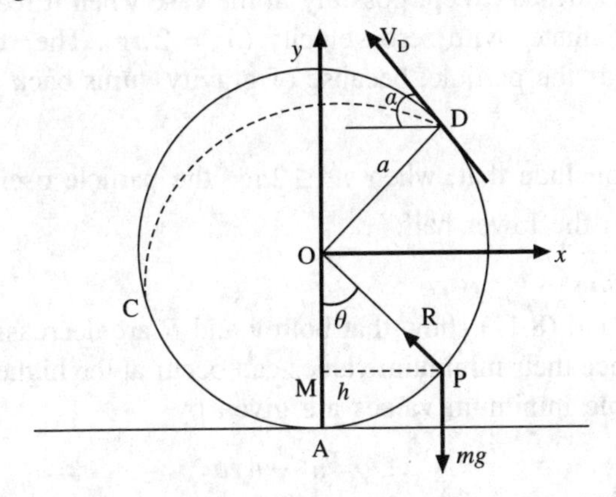

Figure 8.2 The motion of a particle P inside a circle and in case (iii) leaving it at D to pursue a parabolic path.

$$v^2 = u^2 - 2gh, \tag{8.9}$$

where $$h = a(1 - \cos\theta),$$

or $$\cos\theta = \frac{a-h}{a} \tag{8.10}$$

and then from normal equation of motion

$$\frac{v^2}{a} = \frac{R}{m} - g\cos\theta. \tag{8.11}$$

From Eqs. (8.3) and (8.9), we get

$$\frac{R}{m} = \frac{u^2 + ga - 3gh}{a}$$

$$= \frac{u^2 - 2ga + 3ga\cos\theta}{a}. \qquad (8.12)$$

Depending on the value of the initial speed u, there are three cases discussed as follows:

Case (i): $u^2 \le 2ag$

The above inequality taken together with Eq. (8.9) shows that $v = 0$ when $h < a$, i.e., the velocity vanishes somewhere in the lower half of the circular section.

Further Eq. (8.12), written as

$$\frac{R}{m} = \frac{u^2 - 2gh + g(a-h)}{a} = \frac{v^2 + g(a-h)}{a}$$

shows that $R > 0$ except when $v = 0$ and $h = a$, i.e., the particle remains in contact with the surface except possibly in the case when it reaches the end of the horizontal diameter with zero velocity ($u^2 = 2ag$). The velocity vanishes instantaneously as the particle, because of gravity, turns back and resumes its circular path.

Thus, we conclude that, when $u^2 \le 2ag$, the particle oscillates about the mean position in the lower half.

Case (ii): $u^2 \ge 5ag$

Equations (8.9) and (8.12) show that both v and R are decreasing functions of the height h, hence their minimum values can occur at the highest point $h = 2a$. Thus, the possible minimum values are given by

$$v_{\min}^2 = u^2 - 4ga \qquad (8.13)$$

$$\frac{R_{\min}}{m} = \frac{1}{a}(u^2 - 5ga). \qquad (8.14)$$

Equations (8.13) and (8.14) show that if $u^2 > 5ag$, the reaction R_{min} remains positive except possibly at the highest point and that the velocity cannot vanish. Therefore, even when the reaction vanishes at the highest point, the momentum of the particle carries it forward on its circular path. Thus, when $u^2 \ge 5ag$, the particle executes complete circular motion.

Case (iii): $2ag < u^2 < 5ag$

The restriction $u^2 > 2ag$ ensures that the particles do not execute oscillations in lower quadrants and the condition $u^2 < 5ag$ excludes complete circular motion.

Equation (8.12) in conjunction with given inequalities, reveals that R vanishes for certain value of h such that $a < h < 2a$. In other words, the particle leaves the curve at a point D lying in the second quadrant. It is seen that the height h of the point D above the point of projection A is given by

$$h_D = \frac{u^2 + ag}{3}, \tag{8.15}$$

and that the angular displacement θ_D then is given by

$$\cos\theta_D = \frac{2}{3} - \frac{u^2}{3ag}. \tag{8.16}$$

Since $u^2 > 2ag$, Eq. (8.16) shows that $\theta_D > \frac{\pi}{2}$ which may otherwise also be concluded from the fact that D lies in the second quadrant. The particle will leave the curve in a tangential direction making an angle α with the horizontal given by

$$\cos\alpha = \cos(\pi - \theta_D) = \frac{u^2 - 2ag}{3ag} \tag{8.17}$$

with a velocity v_D given by

$$v_D^2 = \frac{u^2 - 2ag}{3}. \tag{8.18}$$

Thus, it is seen that in this case the reaction vanishes and the particle leaves the circular path, the subsequent motion being parabolic with initial velocity v_D and angle of projection α.

Problem 2

A variation of this problem is that of a simple pendulum. Here the particle is tied to one end of a light inextensible string of length l whose upper end is tied to a fixed point; the particle is imparted a horizontal velocity u when at its lowest position in the static situation.

The only difference from Problem 1 is that instead of the constraint force of reaction R we have the constraint force of tension T in the string. The students are advised to write down the details on the lines of the previous problem. We proceed to determine the time for the oscillatory motion corresponding to the Case (i) of Problem 1.

Now, the tangential equation of motion with $s = a\theta$ provides the differential equation

$$\frac{d^2\theta}{dt^2} = -\frac{g}{a}\sin\theta. \tag{8.19}$$

Integrating Eq. (8.19), we obtain

$$\left(\frac{d\theta}{dt}\right)^2 = \frac{2g}{a}(\cos\theta - \cos\alpha), \tag{8.20}$$

where we have used the condition that $\frac{d\theta}{dt} = 0$, when $\theta = \alpha$; α being the maximum angular displacement.

Integrating Eq. (8.20), we obtain the time T for a complete oscillation. It is equal to four times the time taken by the particle to move from $\theta = 0$ to $\theta = \alpha$. Thus, we have

$$T = 4\sqrt{\frac{a}{2g}}\int_0^{\alpha}\frac{d\theta}{\sqrt{(\cos\theta - \cos\alpha)}}. \tag{8.21}$$

The integral given in Eq. (8.21) is elliptic integral and not expressible in closed form in terms of elementary functions. We shall here find an approximation for small values of the displacement α.

Let us write

$$\sqrt{(\cos\theta - \cos\alpha)} = \sqrt{2\left(\sin^2\frac{\alpha}{2} - \sin^2\frac{\theta}{2}\right)},$$

and make the substitution $\sin\left(\frac{\theta}{2}\right) = \sin\left(\frac{\alpha}{2}\right)\sin w$ in the integral on the right hand side of Eq. (8.21). Thus, we have

$$T = 4\sqrt{\frac{a}{g}}\int_0^{\pi/2}\frac{dw}{\sqrt{\left(1 - \sin^2\frac{\alpha}{2}\sin^2 w\right)}} \tag{8.22}$$

$$= 4\sqrt{\frac{a}{g}}\int_0^{\pi/2}\left[1 + \frac{1}{2}\sin^2\frac{\alpha}{2}\sin^2 w + \left(\frac{1}{2}\cdot\frac{3}{4}\right)\sin^4\frac{\alpha}{2}\sin^4 w\right.$$

$$\left. + \left(\frac{1}{2}\cdot\frac{3}{4}\cdot\frac{5}{6}\right)\sin^6\frac{\alpha}{2}\sin^6 w + \cdots + \right] dw$$

$$= 2\pi\sqrt{\frac{a}{g}}\left[1 + \left(\frac{1}{2}\right)^2\sin^2\frac{\alpha}{2} + \left(\frac{1}{2}\cdot\frac{3}{4}\right)^2\sin^4\frac{\alpha}{2} + \left(\frac{1}{2}\cdot\frac{3}{4}\cdot\frac{5}{6}\right)^2\sin^6\frac{\alpha}{2} + \cdots + \right].$$

Therefore, correct up to $0(\alpha^2)$, we have

$$T = T_0\left[1 + \frac{\alpha^2}{16}\right]. \tag{8.23}$$

Here $T_0 = 2\pi\sqrt{\frac{a}{g}}$ is the periodic time of a simple pendulum of small amplitude whose differential equation

$$\frac{d^2\theta}{dt^2} = -\frac{g}{a}\theta$$

follows from Eq.(8.19) by taking the approximation $\sin\theta \approx \theta$.

Problem 3

Motion on the outer side of a smooth surface of circular cross section of radius a.

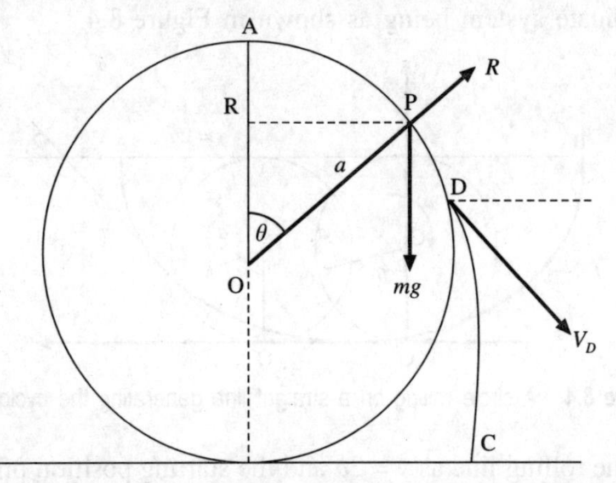

Figure 8.3 Motion on the outer side of a smooth circle the particle P leaving at D to pursue the parabolic path DC.

From Figure 8.3 the particle, being slightly displaced from the highest point A, slides down on the exterior smooth surface. As in Problem 1, using energy equation, we have

$$v^2 = 2gh = 2ag(1 - \cos\theta). \tag{8.24}$$

Next, the normal equation of motion

$$\frac{mv^2}{a} = mg\cos\theta - R$$

on making use of Eq. (8.24) provides

$$\frac{R}{m} = \frac{g}{a}(a - 3h) = g(3\cos\theta - 2). \tag{8.25}$$

Thus, we see that the particle leaves the curve at a point D where

$$h = \frac{a}{3}$$

or

$$\theta = \cos^{-1}\left(\frac{2}{3}\right) \tag{8.26}$$

and the initial velocity of the subsequent parabolic path is given by

$$v_D^2 = \frac{2ga}{3}. \tag{8.27}$$

8.4 MOTION ON A SMOOTH CYCLOID

Cycloid is a curve traced by a point P fixed on the rim of a circle when the latter rolls on a straight line without slipping. Let the rolling circle be of radius a, the co-ordinate system being as shown in Figure 8.4.

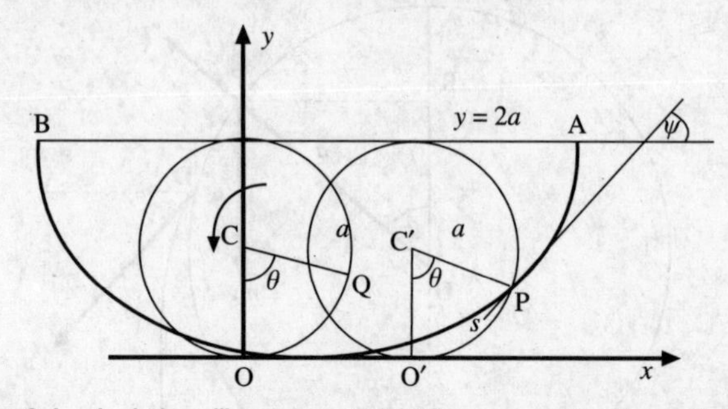

Figure 8.4 A circle rolling on a straight line generating the cycloid AOB.

Taking the rolling line as $y = 2a$ and the starting position of the fixed point at the origin O, the curve traced has the parametric representation

$$x = a(\theta + \sin\,\theta), \tag{8.28}$$

$$y = a(1 - \cos\,\theta). \tag{8.29}$$

The intrinsic equation may be shown to have the form

$$s = 4a\sin\psi, \tag{8.30}$$

where arc distance s is measured from the vertex O and the slope ψ is taken with the x-axis.

Now, we have the radius of curvature

$$\rho = \frac{ds}{d\psi} = 4a\cos\psi. \tag{8.31}$$

Also following relations hold good

$$\psi = 2\theta, \tag{8.32}$$

$$s^2 = 8ay. \tag{8.33}$$

A cusp of the cycloid is at A where $\psi_A = \dfrac{\pi}{2}$ and $s = 4a$; thus whole length of a complete arc of the cycloid is $8a$.

8.4.1 Motion of a Particle on a Cycloid under the Force of Gravity

The tangential and normal equations of motion are given as [refer Figure 8.5]

$$\frac{d^2 s}{dt^2} = -g\sin\psi, \tag{8.34}$$

$$\frac{v^2}{\rho} = \frac{R}{m} - g\cos\psi. \qquad (8.35)$$

Using the intrinsic Eq. (8.30) of the cycloid, Eq. (8.34) becomes

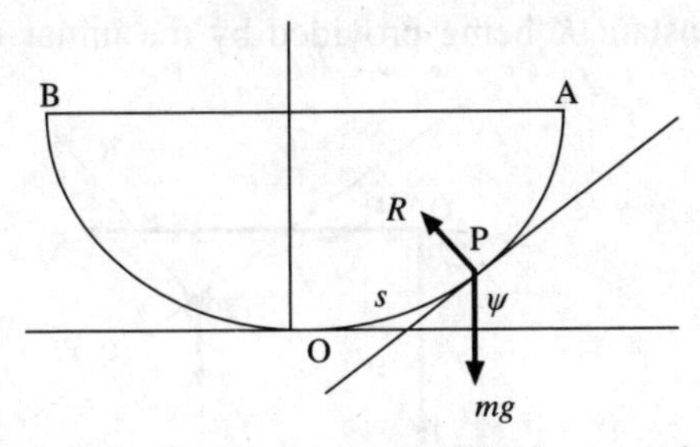

Figure 8.5 Particle P executing SHM on a cycloid.

$$\frac{d^2 s}{dt^2} = -\frac{gs}{4a} \qquad (8.36)$$

which is simple harmonic equation and holds for any value of s which does not take the particle beyond the neighbouring cusps A and B.

Integration of Eq. (8.36) immediately provides the speed v given by

$$v^2 = K - \frac{g}{4a}s^2, \qquad (8.37)$$

where the constant K is to be evaluated from the initial value of v.

Next, Eq. (8.35) provides the reaction R as

$$\frac{R}{m} = \frac{v^2}{4a\cos\psi} + g\cos\psi. \qquad (8.38)$$

It may also be seen that the general solution of Eq. (8.36) is

$$s = \mathrm{A}\cos\sqrt{\frac{g}{4a}}\,t + \mathrm{B}\sin\sqrt{\frac{g}{4a}}\,t, \qquad (8.39)$$

providing

$$v = \frac{ds}{dt} = \sqrt{\frac{g}{4a}}\left[-\mathrm{A}\sin\sqrt{\frac{g}{4a}}\,t + \mathrm{B}\cos\sqrt{\frac{g}{4a}}\,t\right]. \qquad (8.40)$$

The constants A and B are to be determined from the initial conditions.

The time period of a complete oscillation is easily seen to be $4\pi\sqrt{\frac{a}{g}}$.

8.4.2 Motion under Gravity on an Inverted Cycloid

In this case the tangential equation of motion is [refer Figure 8.6]

$$\frac{d^2 s}{dt^2} = g\sin\psi = \frac{gs}{4a}. \qquad (8.41)$$

The first integral of Eq. (8.41) is

$$v^2 = \left(\frac{\mathrm{d}s}{\mathrm{d}t}\right)^2 = K + \frac{gs^2}{4a} \tag{8.42}$$

the value of the constant K being provided by the initial value of v.

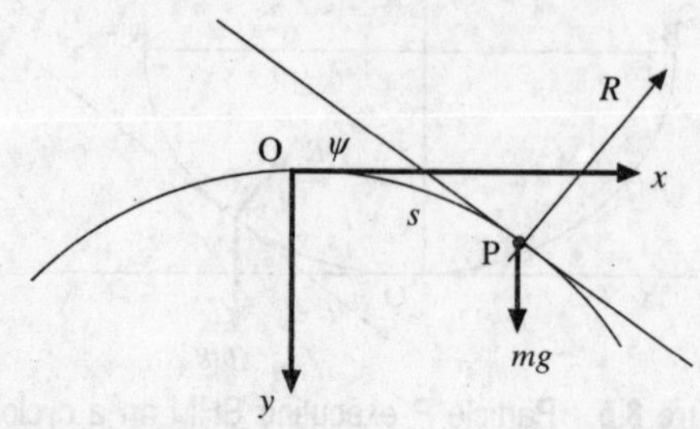

Figure 8.6 Motion of particle P down an inverted cycloid.

Now, from the normal equation of motion, reaction follows as

$$\frac{R}{m} = g\cos\psi - \frac{v^2}{4a\cos^2\psi}. \tag{8.43}$$

In this case, it is seen that the particle leaves the curve when $R = 0$, the subsequent path being parabolic.

8.4.3 Isochronous Pendulum

Simple pendulum provides periodic motion independent of displacement only when the amplitude is small. This difficulty was overcome by Huyghens by designing the isochronous pendulum by exploiting the simple harmonic nature of the motion on a cycloid discussed in Section 8.4.1 and the fact that the evolute (envelope of normals) of cycloid is another equal cycloid.

An isochronous pendulum consists of a small mass m hung from a fixed point C by an inextensible string CDP which is constrained to wind on and off a pair of fixed congruent cycloidal arcs CA and CB having C as a cusp and A and B as vertices [refer Figure 8.7]. As DP, the free portion of the string, is normal to the curve AB (executed by P) and is tangent to the curve CA or the

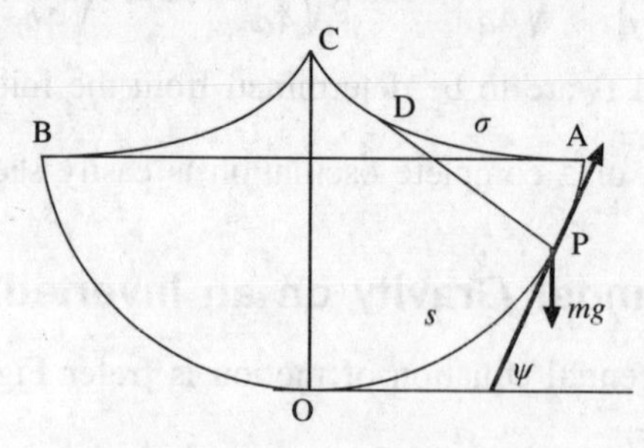

Figure 8.7 Isochronous pendulum; generation of the cycloid AOB.

curve CB, it is concluded that AB is the cycloid whose evolute consists of the cycloidal arcs CA and CB.

Thus, we have a pendulum whose time period $4\pi\sqrt{\dfrac{a}{g}}$ (a being the parameter of the cycloid) is independent of the extent of the amplitude.

We can also determine the equation to the curve AOB by observing that instantaneously the point P is moving in a circle of radius DP with instantaneous centre D.

Thus, we have

$$\frac{\mathrm{d}s}{\mathrm{d}t} = \mathrm{DP}\frac{\mathrm{d}\psi}{\mathrm{d}t} = \sigma\frac{\mathrm{d}\psi}{\mathrm{d}t}, \tag{8.44}$$

where $\sigma = 4a\sin\beta$ is the equation to the cycloid CDA with vertex at A.

Moreover, $\psi = \dfrac{\pi}{2} - \beta$, hence Eq. (8.44) becomes

$$\frac{\mathrm{d}s}{\mathrm{d}t} = 4a\cos\psi\frac{\mathrm{d}\psi}{\mathrm{d}t}$$

which on integration yields

$$s = 4a\sin\psi$$

satisfying the condition that $\psi = 0$ when $s = 0$.

This shows that the curve AOB is a cycloid of the same size as the cycloid CA (or CB).

SOLVED EXAMPLES

EXAMPLE 8.1 A particle slides down the smooth curve $y = c\sinh\left(\dfrac{x}{c}\right)$, the x-axis being horizontal and the y-axis downwards, starting from rest at a point where the tangent is inclined at an angle α to the horizontal; show that it will leave the curve when it has fallen through a vertical distance $c\sec\alpha$.

Solution In Figure 8.8, the given curve is

$$y = c\sinh\left(\frac{x}{c}\right). \tag{i}$$

Therefore,

$$\tan\psi = \frac{\mathrm{d}y}{\mathrm{d}x} = \cosh\frac{x}{c}. \tag{ii}$$

Let A(a, b) be the starting point where $\psi = \alpha$, hence from Eq. (ii)

$$\tan\alpha = \cosh\frac{a}{c} = \sqrt{\left(1 + \frac{b^2}{c^2}\right)}$$

or
$$b^2 = c^2 \tan^2 \alpha - c^2. \tag{iii}$$
Now, from energy equation, speed at any depth y is given by
$$v^2 = 2g(y - b) = 2gh, \tag{iv}$$
where $h = y - b$ is the vertical depth fallen by the particle.

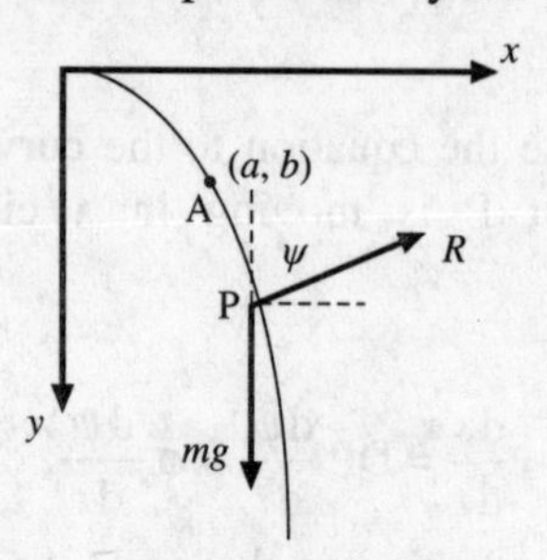

Figure 8.8 Particle P moving down the curve $y = c \sinh (x/c)$.

Next, the normal equation of motion provides
$$\frac{v^2}{\rho} = g \cos\psi - \frac{R}{m}. \tag{v}$$
The particle will leave the curve when $R = 0$, thus, we have from Eq.(v)
$$v^2 = \rho g \cos\psi = g\frac{dx}{d\psi}. \tag{vi}$$
Using Eqs. (i), (ii) and (iv), we obtain from Eq. (vi)
$$2gh = gc\frac{\sec^2\psi}{\sinh\frac{x}{c}} = \frac{gc^2}{y}\left[1 + \cosh^2\frac{x}{c}\right]$$
or
$$2hy = 2c^2 + y^2$$
or
$$2h\,(b + h) = 2c^2 + (b + h)^2$$
or
$$h^2 = c^2 \sec^2 \alpha$$
or
$$h = c \sec\alpha.$$

EXAMPLE 8.2 A particle is free to move on the inner surface of smooth fixed sphere of radius a. The particle is projected horizontally from the lowest point with speed $2\sqrt{ga}$. Show that the particle will leave the sphere after a time $\sqrt{\frac{a}{g}}\log\left(\sqrt{6} + \sqrt{5}\right)$ and that it strikes the sphere again after a further time $\frac{4}{3}\sqrt{\frac{10a}{g}}$.

Solution Here we have $u^2 = 4ag$. Thus, we have $2ag < u^2 < 5ag$ and that is a situation corresponding to the Section 8.3 Case (iii) of Problem 1. Therefore, the particle leaves the circular path at a point D where [See Eq. (8.16)].

$$\cos\theta_D = \frac{2}{3} - \frac{4}{3} = -\frac{2}{3}. \tag{i}$$

Next, integrating the tangential equation of motion once, we get the time in which the particle reaches D as

$$t_D = \sqrt{\frac{a}{2g}} \int_0^{\theta_D} \frac{d\theta}{\sqrt{(1+\cos\theta)}}$$

$$= \sqrt{\frac{a}{2g}} \log\left(\sqrt{6}+\sqrt{5}\right). \tag{ii}$$

The velocity of the particle at D is now [see Eq.(8.19)].

$$v_D = \sqrt{\frac{2ga}{3}}. \tag{iii}$$

Let the particle strike the sphere again at C after a further time t. Then taking the axis at O (refer Figure 8.3), the parametric equation of the parabola provides the co-ordinates of C as

$$x = -a\sin\theta_D + v_D t\cos\theta_D = -\frac{a\sqrt{5}}{3} - \frac{2}{3}\sqrt{\frac{2ga}{3}}\,t,$$

$$y = a\cos\theta_D + v_D t\sin\theta_D - \frac{1}{2}gt^2 = -\frac{2a}{3} + \frac{1}{3}\sqrt{\frac{10ga}{3}}\,t - \frac{1}{2}gt^2.$$

Now, the point C also lies on the circle $x^2 + y^2 = a^2$, and so we have the relation

$$\left\{a\sqrt{5} + 2\sqrt{\frac{2ag}{3}}\,t\right\}^2 + \left\{2a - \sqrt{\frac{10ag}{3}}\,t + \frac{3}{2}gt^2\right\}^2 = 9a^2,$$

giving

$$t = \frac{3}{2}\sqrt{\frac{10a}{g}}.$$

If $u^2 = 7\dfrac{ag}{2}$ show that the particle strikes the spherical surface again at the lowest point.(?)

EXAMPLE 8.3 A particle slides down the surface of a smooth sphere of radius a being slightly displaced from rest at the highest point. Find where it will leave the sphere, and show that it will strike the ground at a distance $5a\dfrac{\left(\sqrt{5}+4\sqrt{2}\right)}{27}$ from the vertical diameter.

Solution From Problem 3, we see that the particle leaves the curve at D with velocity

$$v_D = \sqrt{\frac{ga}{3}},$$

and the angle of the subsequent parabolic path with horizontal at D is $-\theta_D$, where

$$\cos\theta_D = \frac{2}{3}.$$

With axes at O (Figure 8.3), the parabolic path is given by (Refer to the equation of projectile: See Textbook of Dynamics, F. Chorlton)

$$y - a(1+\cos\theta_D) = -(x - a\sin\theta_D)\tan\theta_D - \frac{g}{2v_D^2}(x - a\sin\theta_D)^2 \sec^2\theta_D. \quad \text{(i)}$$

The position of the point C where the particle strikes the ground is obtained by putting $y = 0$ in Eq. (i) and solving for x. Thus, we get

$$x = \frac{5a}{27}\left(\sqrt{5} \pm 3\sqrt{2}\right). \quad \text{(ii)}$$

Omitting the inadmissible −ve sign we have the required solution.

EXAMPLE 8.4 A particle slides down a smooth cycloid starting at rest from the cusp. Find the speed with which it reaches the vertex and the thrust there.

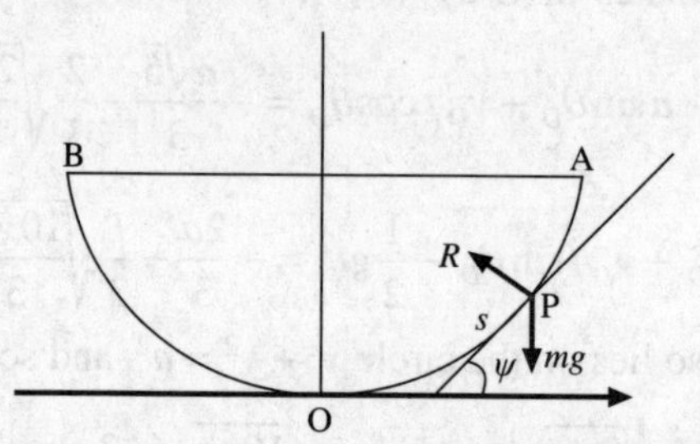

Figure 8.9 Particle P sliding down a smooth cycloid AOB ($s = 4a \sin \psi$).

Solution Keeping in view Figure 8.9, we have [Eq. (8.37)]

$$v^2 = K - \frac{gs^2}{4a}. \quad \text{(i)}$$

Applying the condition $v = 0$ when $s = 4a$, we get $K = 4ag$, and hence Eq. (i) becomes

$$v^2 = \frac{g}{4a}(16a^2 - s^2). \quad \text{(ii)}$$

Further, we know that [Eq. (8.38)]

$$\frac{R}{m} = g\cos\psi + \frac{g(16a^2 - s^2)}{16a^2\cos\psi}. \quad \text{(iii)}$$

Now the value of the speed and the thrust at the vertex are obtained from Eqs. (ii) and (iii) by putting $s = 0$ and $\psi = 0$. We thus have

$$v = 2\sqrt{ag},$$

and

$$R = 2mg.$$

EXAMPLE 8.5 Two particles are let drop from the cusp of a cycloid down the curve at an interval of time t_0; prove that they will meet after a time $2\pi\sqrt{\frac{a}{g}} + \frac{t_0}{2}$.

Solution Suppose the two particles meet at some point D on the left side when the first particle is returning after reaching the end B and the second particle is on its onward journey to B. If t_1 is the time taken to move from B to D and $T = 4\pi\sqrt{\frac{a}{g}}$ is the time of a complete oscillation, then the time taken to move from A to D is $\frac{T}{2} - t_1$.

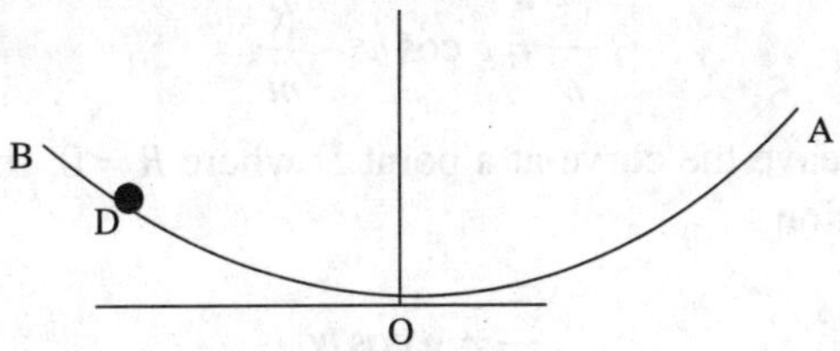

Figure 8.10 The two particle meeting at D on the cycloid AOB.

Thus, we have

T_1 = Time taken by the first particle to reach $D = \frac{T}{2} + t_1$

T_2 = Time taken by the second particle to reach $D = t_0 + \frac{T}{2} - t_1$

Since, two particles meet at D [See Figure 8.10] we have

$$T_1 = T_2$$

or

$$\left(\frac{T}{2}\right) + t_1 = t_0 + \left(\frac{T}{2}\right) - t_1$$

or

$$t_1 = \frac{1}{2}t_0.$$

Hence, the time in which the particles meet is given as

$$\left(\frac{1}{2}\right)T + t_1 = 2\pi\sqrt{\frac{a}{g}} + \frac{1}{2}t_0.$$

EXAMPLE 8.6 A particle starts from the vertex of a cycloid whose axis is vertical and vertex upwards. Prove that it will leave the curve when it has fallen half the vertical height of the cycloid. Also show that the latus rectum of the parabola subsequently described is equal to the height of the cycloid.

Solution With axes as marked in Figure 8.6, the cycloid is

$$s = 4a \sin \psi, \tag{i}$$

providing

$$\rho = \frac{\mathrm{d}s}{\mathrm{d}\psi} = 4a \cos \psi. \tag{ii}$$

Also, we have the standard relation

$$s^2 = 8ay. \tag{iii}$$

Now, from energy equation, we have at a point P at a depth y

$$v^2 = 2gy. \tag{iv}$$

Next, the normal equation of motion provides

$$\frac{v^2}{\rho} = g \cos \psi - \frac{R}{m}. \tag{v}$$

The particle will leave the curve at a point D where $R = 0$; thus, we have from Eq. (v) the condition

$$\frac{v^2}{\rho} = g \cos \psi. \tag{vi}$$

Using Eqs. (i), (ii) and (iv) we get

$$2gy = 4ag \cos^2 \psi = 4ag\left(1 - \frac{y}{2}\right)$$

or

$$y = a. \tag{vii}$$

This shows that the particle will leave the curve when it has fallen half the vertical height ($= 2a$) of the cycloid.

Further the velocity at D is given by

$$v_D^2 = 2gy|_{y=a} = 2ga, \tag{viii}$$

and the angle of inclination ($-\alpha$) of the projectile is obtained from Eqs. (i), (iii) and (vii); we have

$$16a^2 \sin^2 \alpha = 8a^2$$

or

$$\sin \alpha = \frac{1}{\sqrt{2}}$$

or

$$\alpha = \frac{\pi}{4}.$$

Now, the latus rectum of the parabolic path is

$$\frac{2v_D^2 \cos^2 \alpha}{g} = \frac{2 \cdot 2ga}{g} \frac{1}{2} = 2a$$

$$= \text{Height of the cycloid.}$$

PROBLEMS

1. A heavy particle hangs from a point O by a string of length l. It is projected horizontally with a velocity u such that $u^2 = \left(2+\sqrt{3}\right) l g$. Show that the string becomes slack when it has described an angle $\cos^{-1}\left(\frac{-1}{\sqrt{3}}\right)$. Also, show that it will subsequently pass through O.

2. A particle is free to move on a smooth vertical circular wire of radius a. It is projected from the lowest point with velocity just sufficient to carry it to the highest point. Show that the reaction between the particle and the wire is zero after a time

$$\sqrt{\frac{a}{g}} \log\left(\sqrt{5}+\sqrt{6}\right).$$

3. A bead sliding on a fixed smooth vertical circular wire is making complete revolutions. If the ratio of the greatest to the least speed is n: 1 $\left(n < \sqrt{5}\right)$, find the ratio of the greatest to the least force exerted by the wire on the bead. What is the significance of the condition $n < \sqrt{5}$?

[Ans: $(5n^2 - 1)$; $(5 - n^2)$]

4. A heavy particle of mass m is suspended by a string of length l and hangs vertically; it is then projected horizontally with velocity u so that it just makes a complete revolution; show that $u^2 = 5gl$ and that the tension in the lowest position is 6 mg.

5. Prove that, in the case where the pendulum is projected from the position of equilibrium with velocity equal to that due to a fall from the highest point, the time of describing any angle θ is

$$\sqrt{\frac{l}{g}} \log\left(\sec\frac{\theta}{2} + \tan\frac{\theta}{2}\right),$$

where l is the length of the string.

6. A simple pendulum is started so as to make complete revolutions in a vertical plane. If ω_1, ω_2 be the greatest and least angular velocities, prove that the angular velocity when the pendulum makes an angle θ with the vertical is

$$\left(\omega_1^2 \cos^2 \frac{1}{2}\theta + \omega_2^2 \sin^2 \frac{1}{2}\theta\right)^{\frac{1}{2}},$$

and that the tension of the string is

$$T_1 \cos^2 \frac{1}{2}\theta + T_2 \sin^2 \frac{1}{2}\theta,$$

where T_1, T_2 are the greatest and least tensions.

7. A smooth hoop, in the shape of a circle of radius a, is fixed in a vertical plane. A bead, which is threaded on the hoop, passes the highest point of the hoop with a speed $(3ga)^{\frac{1}{2}}$. Find the speed when it passes the lowest point. Show that the force exerted by the hoop on the bead when the bead is at its lowest point is four times the force exerted by the hoop when the bead is at its highest point.

[**Ans:** $(7ga)^{\frac{1}{2}}$]

8. A particle is projected from the lowest point of a smooth sphere of radius a along its inner side with a velocity equal to that it would acquire in falling freely through a height $\frac{7a}{4}$. Show that the particle leaves the sphere after reaching a height $\frac{3a}{2}$ and then returns to the point of projection.

9. A heavy particle, hanging vertically from a fixed point by a light inextensible string of length l, is struck by a horizontal blow which imparts it a velocity $\sqrt{2gl}$; prove that the string becomes slack when the particle has risen to a height $\frac{2l}{3}$ above the fixed point and find the height of the highest point on the parabola subsequently described.

[**Ans:** $\frac{23}{27}l$ above the fixed point]

10. One of the feats of a trick cyclist is to ride inside a sphere, sometimes riding head downwards past the highest point. If the greatest velocity with which he can reach halfway up the sphere is V, and if he does not pedal in the upper hemisphere, what is the diameter of the largest sphere in which he can perform his feat without losing contact with the sphere?

11. Show that a cycloid can be a free path for a particle acted on at each point by a constant force parallel to the corresponding radius of generating circle, this circle being placed at the vertex.

12. A particle is moving on a smooth curve under gravity and its velocity varies as the arcual distance from the highest point. Prove that the curve must be a cycloid.

13. A particle starts from rest at the cusp of a smooth cycloidal arc whose axis is vertical and vertex downwards; prove that, when it has fallen through half the distance measured along the arc to the vertex, two-thirds of the time of descent will have elapsed.

14. A particle falls down a cycloid under its own weight starting from the cusp with velocity u. Find the thrust on the curve at the vertex and the time of reaching the vertex.

$$\left[\textbf{Ans:} \quad 2mg + \frac{mu^2}{4a};\ 2\sqrt{\frac{a}{g}}\tan^{-1}\frac{\sqrt{4ag}}{u}\right]$$

15. A particle is placed near the vertex of a smooth cycloid, the axis being vertical and vertex upwards, find where the particle leaves the curve. Prove that it falls upon the base of the cycloid at a distance $\left(\frac{\pi}{2} + \sqrt{3}\right)a$ from the centre of the base a being the radius of the generating circle.

$$\left[\textbf{Ans:} \quad \psi = \frac{\pi}{4}\right]$$

16. Show that for a particle, sliding down the arc and starting from the cusp of a smooth cycloid whose vertex is lowest, the velocity is maximum when it has described half the vertical height.

17. A particle slides down a catenary, whose plane is vertical and vertex upwards, the velocity at any point being due to the fall from the directrix prove that the pressure at any point varies inversely as the distance of that point from the directrix.

18. A particle descends a smooth curve under the action of gravity, describing equal vertical distances in equal times, and starting in a vertical direction. Show that the curve is a semi-cubical parabola, tangent at the cusp of which is vertical.

19. A wire, in the form of the parabola $y^2 = 4ax$, is fixed with its axis vertical and vertex downwards. If a small smooth bead of mass m can slide on the wire and is released from rest at one end of the latus rectum, find its acceleration along the tangent when it is at a point x above the vertex and show that the thrust on the wire is $2mg\left(\frac{a}{a+x}\right)^{\frac{3}{2}}$.

$$\left[\textbf{Ans:} \quad g\left[\frac{x}{(a+x)}\right]^{\frac{1}{2}}\right]$$

20. A particle moves outside a smooth elliptic cylinder whose axis is horizontal. The major axis of the principal elliptic section is vertical and eccentricity of the section e. If the particle starts from rest on the highest generator, and moves in a vertical plane, it will leave the cylinder at a point whose eccentric angle is β, where $e^2 \cos^3 \beta = 3 \cos \beta - 2$.

21. A particle slides down the curve $x = 2\sqrt{a(y-a)}$ with a velocity due to fall from the x-axis which is horizontal, y-axis being downwards. Find the pressure on the curve at any time, and the time of sliding from $y = b$ to $y = c$.

$$\left[\textbf{Ans:}\quad R = 0;\ \sqrt{\frac{2}{g}}\left\{\sqrt{c-a}-\sqrt{b-a}\right\}\right]$$

22. A particle slides down the smooth curve $y = c\sinh\left(\frac{x}{c}\right)$, the x-axis being horizontal and the y-axis downwards, starting from rest at a point where the tangent is inclined at an angle α to the horizontal; show that it will leave the curve when it has fallen through a vertical distance $c \sec \alpha$.

Orbital Motion

9.1 INTRODUCTION

A number of important problems of practical interest such as Kepler's problem of planetary motion and Rutherford's scattering problem, arise when the external force is a central force — a force which is always directed towards or away from a fixed centre O and whose magnitude is function only of its distance r from the centre of force O. Thus, it may be written as

$$\mathbf{F} = F(r)\,\hat{\mathbf{r}}. \tag{9.1}$$

Physically we have a repulsive or an attractive central force according as $F(r) > 0$ or $F(r) < 0$. In a two body problem, i.e., when two particles interact, keeping one at the origin, the force on the other can be represented as a central force. Gravitational force on planets orbiting round the sun is an example of an attractive force, while the force between nucleus and an alpha particle is a repulsive central force. The most important case is that of the inverse square force $F(r) \propto \dfrac{1}{r^2}$, but other types of forces do occur in more complex interactions of nuclei and molecules.

From the definition contained in Eq. (9.1), we immediately have the line integral round a closed path

$$\oint \mathbf{F} \cdot \mathbf{dr} = 0, \tag{9.2}$$

hence we conclude that central force is conservative.

As Eq. (9.2) stands central force appears to have three components; but this should not lead one to think that the motion under a central force is a screw motion. Actually, it can be shown that the motion under a central force is necessarily planar. To prove it, consider the vector angular momentum $\mathbf{H} = m\,\mathbf{r} \times \mathbf{v}$ of a particle of mass m moving under the central force $\mathbf{F}(\mathrm{r})$. The equation of angular momentum provides

$$\frac{\mathrm{d}\mathbf{H}}{\mathrm{d}t} = \mathbf{r} \times \frac{\mathrm{F}(r)}{r}\mathbf{r} = 0. \tag{9.3}$$

The above result shows that

$$\mathbf{H} = m\mathbf{r} \times \mathbf{v} = \mathbf{H}_0 \tag{9.4}$$

is a constant vector. Therefore, both $\mathbf{r}$ and $\mathbf{v}$ always lie in a fixed plane perpendicular to $\mathbf{H}_0$. We could have inferred this also by observing that, since the force is always directed towards the origin, the particle can never acquire a component velocity out of the plane in which it is initially moving. (What happens when $\mathbf{H}_0$ is a null vector?)

The students are advised to check that Eq. (9.3) follows from the vector equation of motion for the central force, viz.

$$m\frac{\mathrm{d}^2\mathbf{r}}{\mathrm{d}t^2} = \mathrm{F}(r)\,\hat{\mathbf{r}}. \tag{9.5}$$

9.2 EQUATION OF MOTION UNDER A CENTRAL FORCE

The problem, thus, reduces to the consideration of motion in a plane and so is described by two differential equations and four initial conditions. The nature of the central force now warrants that the problem is best tackled in plane polar co-ordinates (r, θ) with the centre of force O as the pole.

We have already learnt about the radial and transverse components of acceleration in Chapter 7. From there, we recall that the equations of motion in the r and θ directions are

$$\text{Radial:} \quad \frac{d^2r}{dt^2} - r\left(\frac{d\theta}{dt}\right)^2 = \frac{F(r)}{m} = -f(r), \tag{9.6}$$

$$\text{Transverse:} \quad \frac{d}{dt}\left(r^2\frac{d\theta}{dt}\right) = 0. \tag{9.7}$$

Here $f(r)$ may be taken as the central force per unit mass taken positive as attractive. The motion is also termed as motion under central acceleration $f(r)$.

Integrating Eq. (9.7) gives

$$r^2\frac{d\theta}{dt} = h, \quad \text{which is a constant} \tag{9.8}$$

whence it can be immediately recognized that the constant h represents the magnitude of the angular momentum $\mathbf{h} = \mathbf{r} \times \mathbf{v}$ [c.f. Eq. (9.4)]. Thus, Eq. (9.8) embodies the conservation of angular momentum about the origin for a motion under a central force.

The principle of conservation of energy is applicable, since central force is conservative, and provides another integral of Eqs. (9.6) and (9.7). We have

$$\text{K.E. is given by} \quad K = \frac{1}{2}v^2 = \frac{1}{2}\dot{r}^2 + \frac{1}{2}r^2\dot{\theta}^2, \tag{9.9}$$

and

$$\text{P.E. is given by} \quad V(r) = \int_{r_0}^{r} f(r)\,dr. \tag{9.10}$$

Thus, the principle of conservation of energy gives

$$\frac{1}{2}v^2 + V = \frac{1}{2}\dot{r}^2 + \frac{1}{2}r^2\dot{\theta}^2 + V = E, \tag{9.11}$$

where E is the energy constant. The student is advised to derive above result starting from the vector equation of motion for central force. Substituting the value of $\frac{d\theta}{dt}$ from Eq. (9.8) in Eq. (9.11), we get

$$\frac{1}{2}\dot{r}^2 + \frac{1}{2}\frac{h^2}{r^2} + V(r) = E. \tag{9.12}$$

We can solve for $\dot{r}$ to get

$$\frac{dr}{dt} = \dot{r} = \sqrt{2\left[E - V(r) - \frac{h^2}{2r^2}\right]}, \tag{9.13}$$

and then integrate to obtain

$$\int_{r_0}^{r} \frac{ds}{\left[E - V(s) - \frac{h^2}{2s^2}\right]^{1/2}} = \sqrt{2}\, t. \tag{9.14}$$

The integral may be evaluated to express r parametrically as a function of t. Using this value in Eq. (9.8), we get on integration

$$\theta = \theta_0 + \int_0^t \frac{h}{r^2}\,dt. \tag{9.15}$$

Thus, we have obtained the solution in terms of the four constants h, E, r_0 and θ_0, which may be evaluated from the initial conditions.

Next, it may be shown that the motion under a central force can always be looked as a rectilinear motion. Thus, we substitute the value of $\frac{d\theta}{dt}$ from Eq. (9.8) in Eq. (9.6) and rewrite the latter as

$$\frac{d^2r}{dt^2} = -f(r) + \frac{h^2}{r^3}, \tag{9.16}$$

which represents the rectilinear motion under the combined effect of the central force $f(r)$ and the fictitious centrifugal force $\frac{h^2}{r^3}$.

Further, the energy Eq. (9.12) may be expressed as

$$\frac{1}{2}\dot{r}^2 + V'(r) = E, \tag{9.17}$$

where

$$V'(r) = \frac{1}{2}\frac{h^2}{r^2} + V(r) \tag{9.18}$$

may be termed as the 'effective potential energy'. Equation (9.17) again embodies the energy principle for a particle moving in a straight line under Eq. (9.16). Some interesting qualitative information about the motion can be deduced from the energy Eq. (9.17) without solving the equations of motion.

Equations (9.14) and (9.15) provide the path parametrically in terms of the parameter t. If the shape of the orbit is needed explicitly as a functional relation between r and θ, using chain rule eliminate t in between Eq. (9.8) and Eq. (9.13) to get

$$h\frac{dr}{d\theta} = r^2\sqrt{\left\{2\left[E - V(r) - \frac{h^2}{2r^2}\right]\right\}} \tag{9.19}$$

giving the orbit

$$\theta = \theta_0 + \int_{r_0}^{r} \frac{dr}{r^2\sqrt{\left\{2\left[E - V(r) - \frac{h^2}{2r^2}\right]\right\}}}. \tag{9.20}$$

Some general results: In polar co-ordinates, we know that the area swept by a radius vector moving from a fixed position r_0 to a variable position r is given by

$$S = \frac{1}{2}\int_{r_0}^{r} r^2\, d\theta. \tag{9.21}$$

This, on using Eq. (9.8), immediately gives the important result

$$\frac{dS}{dt} = \frac{1}{2}h, \tag{9.22}$$

and that may be worded as

For a particle moving under a central force the rate at which area is swept by radius vector is constant equal to half the angular momentum.

If the motion is periodic, then the time period T can be obtained by integrating Eq. (9.22) over a complete period; thus,

$$T = \frac{2S}{h}. \tag{9.23}$$

Further, from our knowledge of calculus we know that the perpendicular from the centre O to the tangent at any point P is given by

$$p = r^2 \frac{d\theta}{ds} = r^2 \frac{d\theta}{dt} / \frac{ds}{dt} = \frac{h}{v}, \tag{9.24}$$

hence, we have $pv = h$

$$v = \frac{h}{p} = h\sqrt{\frac{1}{r^2} + \frac{1}{r^4}\left(\frac{dr}{d\theta}\right)^2}. \tag{9.25}$$

Equation (9.25) is known as Binet's First Formula and may be interpreted as

In a central orbit the speed varies inversely as the perpendicular from the centre upon the tangent to the path.

9.3 DIFFERENTIAL EQUATION OF THE ORBIT

In Eq. (9.20) we have already obtained the orbit in terms of polar co-ordinates (r, θ) by using the energy Eq. (9.11). But it is instructive to derive it directly from the differential equation for the orbit. The equation assumes a simpler form in terms of the variables (u, θ) where $u = \frac{1}{r}$. Thus, the student will find it no difficult to derive, from Eqs. (9.6) and (9.7), the following differential equation

$$\frac{d^2u}{d\theta^2} + u = \frac{1}{h^2u^2} f\left(\frac{1}{u}\right). \tag{9.26}$$

Multiplying both sides of the above equation by $2\frac{du}{d\theta}$ and integrating, we get

$$\left(\frac{du}{d\theta}\right)^2 + u^2 = C + \frac{2}{h^2}\int f\left(\frac{1}{u}\right)\frac{1}{u^2}\,du, \tag{9.27}$$

where C is the constant of integration. The students are advised to work out for themselves that Eq. (9.27) is identical to Eq. (9.19) and to express the constant C in terms of the energy constant E. The orbit then is again given by Eq. (9.20) which when expressed in terms of u assumes the form

$$\theta = \theta_0 - h\int_{u_0}^{u} \frac{du}{\sqrt{2\left[E - V\left(\frac{1}{u}\right) - \frac{1}{2}h^2u^2\right]}}. \tag{9.28}$$

Equation (9.26) can alternately be expressed as

$$f(r) = \frac{h^2}{r^2}\left[\frac{d^2}{d\theta^2}\left(\frac{1}{r}\right) + \frac{1}{r}\right], \tag{9.29}$$

and is then known as *Binet's Second Formula*. It determines the force function $f(r)$ for which the orbit is $r = r(\theta)$.

Now, from the relation

$$\frac{1}{p^2} = \frac{1}{r^2} + \frac{1}{r^4}\left(\frac{dr}{d\theta}\right)^2 = u^2 + \left(\frac{du}{d\theta}\right)^2, \tag{9.30}$$

we can deduce the relation

$$\frac{d^2}{d\theta^2}\left(\frac{1}{r}\right) + \left(\frac{1}{r}\right) = \frac{d^2 u}{d\theta^2} + u = \frac{r^2}{p^3}\cdot\frac{dp}{dr},$$

which when substituted in Eq. (9.29) yields a more convenient formula for determining the force when the orbit is known in pedal form $p = p(r)$; we, thus, have

$$f(r) = \frac{r^2}{p^3}\cdot\frac{dp}{dr}. \tag{9.31}$$

On the other hand, Eq. (9.31) on integration gives the pedal form of the orbit as

$$\frac{1}{p^2} = C - \frac{2}{h^2}\int f(r)dr, \tag{9.32}$$

which, using Eq. (9.30) is found same as Eq. (9.19) or Eq. (9.27).

Thus, we see that the problem of motion under central force has been essentially reduced to solving the second order ordinary differential equation namely Eq. (9.26). Its solution in general will involve two arbitrary constants which are determined by the use of appropriate conditions. These conditions are usually prescribed at points called *Apses*. An apse is a point in a central orbit at which the radial distance drawn from the centre of force has a maximum or minimum value. Thus at an apse $\frac{dr}{d\theta}$ vanishes, consequently, from Eq. (9.30) $p = r$ there. Moreover, it is not difficult to see that at an apse the radial velocity vanishes and the particle moves at right angles to the radius vector with velocity $v = \frac{h}{r}$. The angle between any two consecutive apsidal distances is called *apsidal angle*.

In the case of planetary motion the positions of the least and the greatest distances from the Sun are called *perihelion* and *aphelion*. For a satellite orbiting the Earth, these are known as *perigee* and *apogee*.

9.4 PLANETARY MOTION

We know that the motion of planets round the sun or the motion of the moon or of an artificial satellite about the earth can be modelled by a point mass orbiting under a central force which following Newton's Gravitational law is the inverse square law

$$F(r) = -G\frac{mM}{r^2}. \tag{9.33}$$

Here G is the gravitational constant, m the mass of the orbiting body P, M the mass of the attracting body supposed fixed at O in an inertial frame and r the polar distance OP.

Earlier we have reduced a central force problem to solving a second order differential equation for the orbit which in the present case of inverse square force becomes

$$\frac{d^2u}{d\theta^2} + u = \frac{\mu}{h^2}. \tag{9.34}$$

The general solution of Eq. (9.34) is easily found to be

$$\frac{L}{r} = Lu = 1 + e\cos(\theta - \alpha), \tag{9.35}$$

where $L = \dfrac{h^2}{\mu}$, and e and α are the two constants of integration. The students will easily recognize Eq. (9.35) as the equation to a conic section with semi-latus-rectum L and eccentricity e. We can conveniently choose the initial line and set $\alpha = 0$, thereby getting the simplified equation

$$\frac{L}{r} = Lu = 1 + e\cos\theta. \tag{9.35a}$$

9.4.1 Pedal Form

The first integral of Eq. (9.34) is easily seen to be

$$\left(\frac{du}{d\theta}\right)^2 + u^2 = \frac{\mu}{h^2}\left(\frac{2}{r} + K\right), \tag{9.36}$$

where K is the constant of integration. Now, using the well known relation

$$\frac{1}{p^2} = u^2 + \left(\frac{du}{d\theta}\right)^2,$$

we obtain the pedal form of the orbit as

$$\frac{h^2}{p^2} = \mu\left(\frac{2}{r} + K\right). \tag{9.37}$$

We again recognize Eq. (9.37) as the pedal equation of conic section which will be an ellipse, parabola or hyperbola according as $K < 0$, $K = 0$ or $K > 0$. It should also be noticed that the nearer branch of the hyperbola is being described in this case unlike the description of the further branch in the case of Rutherford scattering (Section 9.6.1). Equation (9.37) may be put in the standard form

$$\text{Ellipse:} \quad \frac{b^2}{p^2} = \frac{2a}{r} - 1, \quad \left(K = -\frac{1}{a} < 0\right) \tag{9.38}$$

$$\text{Parabola:} \quad p^2 = \frac{1}{2}Lr, \quad (K = 0) \tag{9.38a}$$

$$\text{Hyperbola:} \quad \frac{b^2}{p^2} = \frac{2a}{r} + 1, \quad \left(K = \frac{1}{a} > 0\right), \tag{9.38b}$$

where a is the semi-major axis and $b = \sqrt{La} = h\sqrt{\dfrac{a}{\mu}}$ the semi minor axis. It may also be noted that for an ellipse $L = a(1 - e^2)$ and for a hyperbola $L = a(e^2 - 1)$.

From Binet's first formula [see Eq. (9.39) and Eq. (9.36), we get

$$v^2 = h^2\left[\left(\frac{du}{d\theta}\right)^2 + u^2\right] = \mu\left(\frac{2}{r} + K\right). \tag{9.39}$$

It is seen that the velocity at a point depends on the polar distance and not on the direction of motion.

We may also show using Eqs. (9.38a) and (9.39) that if a particle is projected at a distance r_0 from the focus with velocity v_0 then the path is an ellipse, a parabola or a hyperbola given by

$$v_0^2 <=> V_1^2, \tag{9.40}$$

where
$$V_1^2 = -2\mu\int_{\infty}^{r_0} \frac{dr}{r^2} = \frac{2\mu}{r_0}. \tag{9.41}$$

V_1 above is the velocity acquired in falling from infinity to the field position r_0 under the given central force. It may also be interpreted as the escape velocity for a particle starting initially from a distance r_0 and moving under the attraction of the inverse square force.

Using the values of K as given in Eq. (9.38), we may express Eq. (9.39) as

$$v^2 = \mu\left(\frac{2}{r} - \frac{1 - e^2}{L}\right), \tag{9.42}$$

where $e < 1$, $e = 1$ and $e > 1$ naturally correspond respectively to elliptic, parabolic and hyperbolic orbits. We present below a number of useful results.

Time of describing an arc

The time of describing an arc follows from the angular momentum equation; thus, we have

$$t = \frac{1}{h}\int_{\alpha}^{\theta} r^2 d\theta$$

$$= \frac{L^2}{h}\int_{\alpha}^{\theta} \frac{d\theta}{(1+e\cos\theta)^2}. \tag{9.43}$$

Taking the initial position $\alpha = 0$, the student is advised to work out following values of t.

Elliptic orbit

$$t = \frac{a^{\frac{3}{2}}}{\sqrt{\mu}}\left[2\tan^{-1}\left(\sqrt{\frac{1-e}{1+e}}\tan\frac{\theta}{2}\right) - e\sqrt{(1-e^2)}\frac{\sin\theta}{1+e\cos\theta}\right] \tag{9.44}$$

Parabolic orbit

$$t = \frac{L^{\frac{3}{2}}}{2\sqrt{\mu}}\left[\tan\frac{\theta}{2} + \frac{1}{3}\tan^3\frac{\theta}{2}\right] \tag{9.45}$$

Hyperbolic orbit

$$t = \frac{a^{\frac{3}{2}}}{\sqrt{\mu}}\left[e\sqrt{(e^2-1)}\frac{\sin\theta}{1+e\cos\theta} - \log\frac{\sqrt{1+e}+\sqrt{e-1}\tan\frac{\theta}{2}}{\sqrt{1+e}-\sqrt{e-1}\tan\frac{\theta}{2}}\right] \tag{9.46}$$

The radial velocity in an elliptic orbit may be expressed as

$$v_r^2 = \left(\frac{dr}{dt}\right)^2 = \frac{2\mu(r-r_1)(r_2-r)}{r^2(r_1+r_2)}, \tag{9.47}$$

where

$$r_1 = L/(1+e) = a(1-e),$$

$$r_2 = L/(1-e) = a(1+e), \tag{9.48}$$

are apsidal distances.

When the polar distances of apses r_1 and r_2 are known, the value of t for elliptic orbit follows immediately from Eq. (9.47) as

$$t = \frac{\sqrt{r_1+r_2}}{\sqrt{2\mu}}\int\frac{r\,dr}{\sqrt{(r-r_1)(r_2-r)}}. \tag{9.49}$$

The corresponding formula for a hyperbolic orbit is

$$t = \frac{\sqrt{r_2 - r_1}}{\sqrt{2\mu}} \int \frac{rdr}{\sqrt{(r - r_1)(r_2 + r)}}, \tag{9.50}$$

where now $r_1 = \dfrac{L}{(e+1)}$ and $r_2 = \dfrac{L}{(e-1)}$. The students are advised to derive the result for a parabolic orbit and compare it with Eq. (9.45). The periodic time for describing the closed elliptic path may be obtained from the formula given in Eq. (9.44), but it immediately follows from formula given in Eq. (9.23). Thus, we have periodic time

$$T = \frac{2\pi\, ab}{h} = \frac{2\pi\, a^{\frac{3}{2}}}{\sqrt{\mu}} = \frac{\pi(r_1 + r_2)^{\frac{3}{2}}}{\sqrt{2\mu}}. \tag{9.51}$$

Notice that the quantity μ is same for all bodies orbiting round the same centre of attraction.

Students should find it no difficult to show that for a particle projected with velocity v_0 from a point at a distance r_0 from the centre of force, the periodic time can be put in the form

$$T = \frac{2\pi}{\sqrt{\mu}} \left[\frac{2}{r_0} - \frac{{v_0}^2}{\mu} \right]^{\frac{-3}{2}}, \tag{9.52}$$

also that

$$T = a \int \frac{dt}{r}, \tag{9.53}$$

where integration is taken over a complete planetary year.

Average kinetic energy

The mean or average kinetic energy of a particle moving in a closed circuit is defined as

$$\bar{K} = \frac{1}{2T} \int_0^T v^2 dt, \tag{9.54}$$

where T is the time period of orbital motion. Now, inserting the value of v^2 from Eq. (9.42) in Eq. (9.54) and exploiting the following known results

$$r^2 \frac{d\theta}{dt} = h, \ \frac{L}{r} = 1 + e\cos\theta, \ L = \frac{h^2}{\mu},$$

we obtain for an elliptic orbit

$$\bar{K} = \frac{h}{2T} \int_0^{2\pi} \frac{1 + 2e\cos\theta + e^2}{(1 + e\cos\theta)^2} d\theta$$

$$= \frac{\pi h}{T\sqrt{1-e^2}} = \frac{1}{2}\left(\frac{h}{b}\right)^2, \tag{9.55}$$

where Eq. (9.51) has been used. The student should convince himself that Eq. (9.55) embodies the statement.

Average kinetic energy of a planet is equal to the kinetic energy at the ends of the minor axis of the orbit.

He should also show that the total energy E of a planet is the geometric mean of its kinetic energies K_1 and K_2 at perihelion and aphelion, i.e.

$$E^2 = K_1 K_2, \tag{9.56}$$

and further that the average potential energy V is twice the mean kinetic energy

$$\overline{V} = -2\overline{K}. \tag{9.57}$$

9.5 KEPLER'S LAWS

From the foregoing analysis, the discerning student can glean out the following facts:

(I) The path of a planet is an ellipse with the sun at one of its foci.
(II) The area swept out by the radius vector joining the planet to the sun is constant,
(III) The square of the periodic time is proportional to the cube of the mean distance from the sun.

These are natural fall outs of deploying Newton's laws of motion in conjunction with his gravitational law. For a justification of this statement, the student should have a look at Eqs. (9.35a) (Law I), (9.35) (Law II) and (9.51) (Law III). But they were first advanced by the German astronomer Johannes Kepler (1571–1630) on the basis of analyzing extensive observations recorded earlier by his teacher, the celebrated Dane astronomer Tycho Brahe (1546–1601) who believed in the erroneous theory of the Greek astronomer Claudius Ptolemy (AD 127–145). According to Ptolemaic theory, the sun and planets were supposed to orbit the stationary earth, considered as the centre of the Universe, in epicycles — that is, small circles whose centres move around the circumference of larger circles. Ptolemy advanced a complex and cumbersome system to explain the irregular behaviour of planets and attempted to justify his theory by feeding fabricated data. For more than thirteen centuries the wrong notions of Ptolemy ruled not only the astronomical world but the entire Christian religious establishments until over thrown by the father of modern astronomy, the polish astronomer Nicolaus Copernicus (1473–1543). He proposed the heliocentric theory that the sun is the centre of the entire Universe and that the planets are all of the same size and that they move round the sun in circular orbits. Kepler realized that Copernicus theory was only partly true in that the earth and the planets orbit round the sun. He refined the theory and formulated the three important laws of planetary motion. They are known as Kepler's laws of planetary motion.

The first law tells that planets do not move in circular orbits as thought by Copernicus, but in ellipses with the sun occupying one of the foci. The second law implies that closer a planet approaches the sun, the faster it moves in its orbit. This implies that the maximum and minimum velocities occur respectively at the perihelion and aphelion, these being expressible as $2\pi\dfrac{ar_2}{bT}$ and $2\pi\dfrac{ar_1}{bT}$, r_1 and r_2 being perihelion and aphelion distances. His third law gives a simple relationship between the time taken by a planet to complete an orbit round the sun and its mean distance from the sun.

Kepler knew that the sun was exerting a strong influence on planetary motion and thought that some kind of magnetic force linked them. But he was unable to explain either the nature or likely origin of the force. It was left to the genius of Isaac Newton (1642–1727), nearly fifty years later, to propound the universal law of gravitation and to demonstrate that Kepler's laws were a natural consequence of the gravitational force and laws of motion.

9.6 RUNGE-LENZ VECTOR AND RUTHERFORD SCATTERING

Let us consider the vector defined by

$$\mathbf{A} = \mathbf{v} \times (\mathbf{r} \times \mathbf{v}) - r^2 f(r)\hat{\mathbf{r}}, \tag{9.58}$$

where $\mathbf{v} = \dfrac{d\mathbf{r}}{dt}$ is the velocity vector. Observing that $\mathbf{h} = \mathbf{r} \times \mathbf{v}$ is a constant vector for motion under central force system being studied here, and making use of the equation of motion

$$\frac{d\mathbf{v}}{dt} = -f(r)\hat{\mathbf{r}}$$

we find that

$$\frac{d\mathbf{A}}{dt} = -\frac{d(r^2 f)}{dt}\hat{\mathbf{r}} = 0. \tag{9.59}$$

Above result shows that where

$$f(r) = \frac{\mu}{r^2}, \tag{9.60}$$

i.e., for motion under Inverse Square Law $\dfrac{d\mathbf{A}}{dt} = 0$. This implies that the vector

$$\mathbf{A} = \mathbf{v} \times (\mathbf{r} \times \mathbf{v}) - \mu\hat{\mathbf{r}}, \tag{9.61}$$

is a constant vector. The vector **A** defined by Eq. (9.61) is known as Runge-Lenz vector. It will be seen to conveniently describe the motion subject to a central force obeying inverse square law. Analogues of this vector have been earlier employed by Laplace and by Hamilton to discuss the planetary motion.

Using the results of foregoing analysis, the magnitude $A = |\mathbf{A}|$ of the Runge-Lenz vector can be shown to be given by

$$A^2 = 2h^2 E + \mu^2. \tag{9.62}$$

The students are advised to determine the fixed direction of the constant vector **A**. Now, taking dot product of each side of Eq. (9.61) with **r**, we get

$$\mathbf{r} \cdot \mathbf{A} = (\mathbf{r} \times \mathbf{v}) \cdot (\mathbf{r} \times \mathbf{v}) - \mu \mathbf{r} \cdot \hat{\mathbf{r}}$$

or

$$rA\cos\theta = h^2 - \mu r, \tag{9.63}$$

where θ is the angle between the directions of **r** and **A**. Equation (9.63) may be recast as

$$\frac{L}{r} = 1 + e\cos\theta, \tag{9.64}$$

where

$$L = \frac{h^2}{\mu} \text{ and } e = \frac{A}{\mu} = \left(1 + \frac{2h^2 E}{\mu^2}\right)^{\frac{1}{2}}. \tag{9.65}$$

Equation (9.64) is the familiar polar equation of a conic section (ellipse, parabola, hyperbola) with eccentricity e and semi-latus rectum L. Thus, we see that the orbit of a particle moving under a central force obeying inverse square law, is a conic section whose nature is determined by the sign of energy E as follows

$$\begin{aligned} &E > 0,\ e > 1 \text{ Hyperbola,} \\ &E = 0,\ e = 1 \text{ Parabola,} \\ &E < 0,\ e < 1 \text{ Ellipse,} \\ &E = -\frac{\mu^2}{2h^2},\ e = 0 \text{ circle.} \end{aligned} \tag{9.66}$$

9.6.1 Rutherford's Scattering Problem

Rutherford used a beam of α-particles to investigate the structure of atoms. The α-particle corresponds to a helium nucleus consisting of two protons and two neutrons. Thus, if q is the charge on a proton, the charge on an alpha particle is $2q$. The charge on the nucleus of an atom of atomic number N is Nq; hence, the central force per unit mass on the α-particle as given by Coulomb's law is

$$f(r) = -\frac{2Nq^2}{4\pi\,\varepsilon_0 m r^2} = -\frac{\lambda}{r^2}, \tag{9.67}$$

where the constant ε_0 is the permittivity of the free space, m is the mass of the α-particle and

$$\lambda = \frac{Nq^2}{4\pi\,\varepsilon_0 m}, \tag{9.68}$$

a positive constant. Comparing Eq. (9.67) with Eq. (9.60), we see that $\mu = -\lambda$; the minus sign indicates that the force is repulsive rather than an attractive one.

Now, with $f(r)$ given by Eq. (9.67), the P.E. becomes

$$V(r) = -\lambda \int_{\infty}^{r} \frac{dr}{r^2} = \frac{\lambda}{r}, \tag{9.69}$$

and so the energy Eq. (9.11) gives

$$E = \frac{1}{2}v^2 + \frac{\lambda}{r}. \tag{9.70}$$

Since $\lambda > 0$, we see that $E > 0$; hence, by criterion given in Eq. (9.66), the path is hyperbolic, the further branch being described as the latus rectum $2L = \dfrac{-h^2}{\lambda}$ [Eq. 9.65)] is negative.

9.7 MOTION OF AN ARTIFICIAL SATELLITE

The age of space flight began with the historical launch of earth's first artificial satellite *Sputnik*, on October 4, 1957, in USSR. This great event was followed by the first round the earth flight of the Soviet Cosmonaut Yu.A. Gagarin on April 12, 1961, aboard Vostok 1, and by the first foot print of a human being (U.S. Astronaut Neil Armstrong) on the moon, on July 21, 1969. At present, there are many artificial satellites orbiting round the earth and quite a few space ships launched from the earth have become satellites of the sun. The study of the motion of artificial satellites and space ships has developed into a vast subject that basically depends on the orbit theory presented earlier.

A space ship launched from the earth is accelerated into the orbit by rockets which burn their fuel at great speed and so act only for a short period of time. After that, the space ship orbits in free flight. Thus, the path of the space ship consists of two legs: the active leg trajectory in which the engine is active, and the inactive leg in which the engine is inactive. We have simplified the problem by neglecting the initial few minutes of the flight, i.e. we assume that the rocket accelerates the space ship instantaneously to its orbital energy E. Once in orbit the space ship or the artificial satellite will also be influenced by other celestial bodies, viz. the sun, the moon, the planets, etc. But when its distance from the earth is not very large, the force is primarily that of the earth's gravitational field, and if the orbital energy E is negative it becomes an artificial satellite of the earth. Further, since the mass of the space craft is negligible as compared to that of the earth, the centre of the mass of the earth and the space craft system can be taken at the geocentre, and we have the satellite moving relative to an inertial frame of reference fixed in the earth. But it must not be forgotten that actually the earth and the space craft system form a non-inertial system relative to the sun.

In the first approximation, the earth can be regarded as spherical, its gravitational field obeying inverse square law with the centre of force at the geocentre. In this case $\mu = G M_E$, where M_E is the mass of the earth. The value of μ may be expressed in terms of g_0, the value of the acceleration due to gravity at the surface of the earth which is being considered as a sphere of radius r_0. Thus,

$$\mu = GM_E = g_0 r_0^{\,2}. \tag{9.71}$$

Cosmic velocities

From Kepler's second law [Eq. (9.42)], we find that the maximum velocity occurs at the perihelion, i.e. at the nearer end of the major axis and is given by

$$v_1^2 = \mu\left(\frac{2}{r_1} - \frac{1}{a}\right) = 2g_0 r_0^2\left(\frac{1}{r_1} - \frac{1}{r_1 + r_2}\right) \tag{9.72}$$

r_1 and r_2 being perigee and apogee distances.
Equation 9.72 shows that amongst all orbits having the same value of r_1, v_1 will be least when r_2 has the least possible value which is r_1; then the path is circular. This provides the least velocity of projection to put a satellite in orbit around the earth

$$V_0 = r_0\sqrt{\frac{g_0}{r_0 + z}}, \tag{9.73}$$

where $r_1 = r_0 + z$, z being the height above the surface of the earth of the circular path in which the satellite is orbiting. This velocity is called circular velocity or the orbital velocity or the critical velocity. Since, the satellite moves near the surface of the earth, $z << r_0$, and then we have the approximation

$$V_0 \approx \sqrt{g_0 r_0} = \sqrt{9.8 \times 6.4} \simeq 7.9 \text{ km/s}, \tag{9.74}$$

where $g = 9.8 \text{ m/s}^2$ is the value of the acceleration due to gravity and $r_0 = 6.4 \times 10^3$ m is the radius of the earth. For launch velocity less than V_0 the elliptical path will not completely surround the earth and the satellite will fall down on it. The student should satisfy himself that then the arc of the elliptic path will approximate the well known parabolic path of a projectile moving under gravity. The velocity $V_0 = 7.93$ km/s is known as the *first cosmic velocity*. It is the smallest velocity which should be imparted to a body at a geocentric distance equal to the radius of the earth for the body to become an artificial satellite of the earth and the orbit then is circular.

We have already encountered the second cosmic velocity

$$V_1 = \sqrt{\frac{2\mu}{r_0 + z}} \simeq \sqrt{2g_0 r_0} \simeq 11.2 \text{ km/s}. \tag{9.75}$$

It is the escape velocity, being the smallest velocity necessary for launching a space craft on the Earth's surface so that it leaves earth's attraction and becomes a satellite of the Sun. Equation (9.42) immediately provides this value on realizing that for achieving escape the path must at least be parabolic ($e = 1$); hence, it is also referred to as *Parabolic velocity*.

For initial velocity greater than V_1, the path is hyperbolic ($e > 1$) and the space craft again escapes the earth's gravitational field. Thus, as observed earlier, it is the minimum launching velocity to take the space craft beyond the earth's gravitational field. The relation between V_0 and V_1 is

$$V_1 = \sqrt{2}\, V_0. \tag{9.76}$$

When the space craft is far away from the earth's gravitational field, the attraction exerted on it by the sun, by the moon and by other planets also comes into play and modifies the inverse square law. It may also be remarked that variation in this law also occurs because of the non-uniformity of the earth's structure and its deviation from the assumed spherical shape.

The minimum initial velocity required by a space ship, when launched from the earth's surface, to leave the solar system is called the *solar escape velocity* or the *third cosmic velocity*. More complicated calculations show that it is approximately 16.7 km/s.

A little reflection on the part of the students shall tell them which one of the following assertions is true:

The escape velocity of a satellite is:

(i) Greatest for a horizontal launching
(ii) Least for a vertical launching
(iii) Greatest for a vertical launching
(iv) None of these.

The *heading angle* β is defined as the angle that the direction of the motion makes with the perpendicular to the radius vector. Suppose, we have the values

$$\beta = \beta_0, \; v = v_0, \; r = r_0,$$

then exploiting the results derived previously, we can derive the results

$$e^2 = 1 - \frac{r_0^{\,3} v_0^{\,2} \cos^2 \beta_0 (2\mu - r_0 v_0^{\,2})}{\mu^2} \tag{9.77}$$

$$\cos\theta_0 = \frac{\left(\dfrac{r_0 v_0^{\,2}}{\mu} \cos^2 \beta_0 - 1\right)}{e}, \tag{9.78}$$

providing the values of the eccentricity of the orbit and the angular position where the satellite enters the orbit.

9.8 SATELLITE IN CIRCULAR ORBITS

We have seen that if the satellite is launched with orbital velocity V_0, the path is a circle. We shall discuss here the motion in a circular orbit afresh. In a frame of reference fixed on it the satellite is in equilibrium under the action of the inward gravitational force given by $-\left(\frac{GmM_E}{r^2}\right)\hat{\mathbf{r}}$ and the outward fictitious centrifugal force $mr\omega^2\hat{\mathbf{r}}$, ω being the uniform angular velocity of the satellite, with geocentre as the origin. Thus, we have

$$mr\omega^2\hat{\mathbf{r}} - \frac{GmM_E}{r^2}\hat{\mathbf{r}} = 0,$$

giving

$$\omega^2 = \frac{GM_E}{r^3} = \frac{g_0 r_0^2}{(r_0 + z)^3}. \tag{9.79}$$

The time period for a circular orbit, now, can be written as

$$T = \frac{2\pi}{\omega} = \frac{2\pi(r_0 + z)^{\frac{3}{2}}}{r_0\sqrt{g_0}}. \tag{9.80}$$

For small values of z, we then have

$$T \approx 2\pi\sqrt{\frac{r_0}{g_0}} = \frac{2\pi}{60}\sqrt{\frac{6.4\times 10^6}{9.8}} \approx 83 \text{ min}. \tag{9.81}$$

9.8.1 Syncom Satellites

These are communication satellites whose orbits are such that they appear stationary to an observer fixed on the earth. These orbits are therefore circular lying in the equatorial plane, and the rotation of the satellite synchronizes with that of the earth. This implies that the angular velocity ω corresponds to one rotation per day, i.e,

$$\omega = \frac{2\pi}{24\times 60\times 60} \text{ rad/s}. \tag{9.82}$$

The radius of the orbit in which the Syncom satellite moves round the earth is now given by Eq. (9.79)

$$r = \left(\frac{GM_E}{\omega^2}\right)^{\frac{1}{3}} = \left(\frac{g_0 r_0^2}{\omega^2}\right)^{\frac{1}{3}}. \tag{9.83}$$

Substituting the appropriate values, the student should verify that the radius of the Syncom satellite orbit is approximately 4.23×10^4 km, i.e. about seven times the radius of the earth. Also, the height of the satellite above the surface of the earth is

$$z = r - r_0 = 3.59 \times 10^4 \text{ km}. \tag{9.84}$$

Now, the angle (2α) subtended by the base of the enveloping cone of the satellite at the centre of the earth is given by

$$\cos\alpha = \frac{r_0}{r} = \frac{6.4}{42.3}. \tag{9.85}$$

This provides the value

$$2\alpha \approx 162°. \tag{9.86}$$

Thus, one satellite covers an angle of 162° and at least three Syncom satellites are needed for complete coverage of all points on the equator.

SOLVED EXAMPLES

EXAMPLE 9.1 Find the equation of the orbit of a particle if the central force is $\frac{\mu}{r^5}$ per unit mass and the speed of the particle at an apse at a distance of 1 unit is $\sqrt{\frac{\mu}{2}}$.

Solution Let us solve the orbital differential equation

$$\frac{d^2u}{d\theta^2} + u = \frac{\mu u^5}{h^2u^2}. \tag{i}$$

Under the conditions at the apse

$$p = r = 1,\ v = \sqrt{\frac{\mu}{2}},\ \frac{du}{d\theta} = 0. \tag{ii}$$

Now the relation $v = \frac{h}{p}$ under the first of above conditions provides the value $\mu = 2h^2$, and hence Eq. (i) becomes

$$\frac{d^2u}{d\theta^2} + u = 2u^3,$$

which on integration gives

$$\left(\frac{du}{d\theta}\right)^2 = u^4 - u^2 \tag{iii}$$

the constant of integration vanishing on using Eq. (ii).

Expressing Eq. (iii) in terms of $r = \frac{1}{u}$ and taking the square root we obtain

$$\frac{dr}{d\theta} = \sqrt{1 - r^2}. \tag{iv}$$

On integration, Eq. (iv) yields the orbit

$$r = \cos\theta,$$

when we take $\theta = 0$ at $r = 1$.

EXAMPLE 9.2 A particle is moving under a central force along the curve $r = Ce^{-2\theta}$. Show that the force is proportional to $\frac{1}{r^3}$.

Solution We are given

$$r = Ce^{-2\theta} \tag{i}$$

or

$$u = \frac{1}{C}e^{2\theta}. \tag{ii}$$

Now the force is given by Binet's second formula, namely

$$f(r) = \frac{h^2}{r^2}\left[\frac{d^2}{d\theta^2}\left(\frac{1}{r}\right) + \frac{1}{r}\right] = h^2u^2\left[\frac{d^2u}{d\theta^2} + u\right]$$

$$= \frac{5h^2u^2}{C}e^{2\theta} = \frac{5h^2}{r^3}.$$

the last two steps being written down by making use of Eqs. (ii) and (i).

EXAMPLE 9.3 When a particle describes an ellipse about a centre of force at the focus; show that the velocity can be resolved into two components of constant magnitude, one perpendicular to the major axis and the other perpendicular to the radius vector.

Hint

$$\mathbf{v} = \frac{h}{p}\hat{\mathbf{t}} = \frac{h}{p}\frac{d\mathbf{r}}{ds} = \frac{h}{p}\left(\frac{dr}{ds}\hat{\mathbf{r}} + r\frac{d\theta}{ds}\hat{\boldsymbol{\theta}}\right)$$

$$= \frac{h}{r\sin\varphi}\left[\frac{(\hat{\mathbf{j}} - \hat{\boldsymbol{\theta}}\cos\theta)}{\sin\theta}\cos\varphi + \sin\varphi\hat{\boldsymbol{\theta}}\right]$$

$$= \frac{h}{r}\left[\frac{\cot\varphi}{\sin\theta}\hat{\mathbf{j}} - (\cot\theta\cot\varphi - 1)\hat{\boldsymbol{\theta}}\right].$$

Now use the polar equation $\frac{L}{r} = 1 + e\cos\theta$ to show that the coefficients of $\hat{\mathbf{j}}$ and $\hat{\boldsymbol{\theta}}$ are constants.

EXAMPLE 9.4 Show that the time in a central orbit whose pedal equation is given by

$$t = \frac{1}{h}\int\frac{prdr}{\sqrt{r^2 - p^2}},$$

where h is the constant angular momentum.

Solution The radial velocity may be expressed as

$$\frac{dr}{dt} = v\cos\varphi$$

$$= \frac{h}{p}\sqrt{1-\frac{p^2}{r^2}}, \qquad \text{(i)}$$

as $v = \dfrac{h}{p}$ and $p = r\sin\varphi$.

Next separating the variables and integrating we get the required result.

EXAMPLE 9.5 A particle is projected at a distance r_0 from the focus with velocity v_0, show that the path will be a rectangular hyperbola if the angle of projection is $\sin^{-1}\left[\dfrac{\mu}{\left\{v_0 r_0 (v_0^2 - V_1^2)^{\frac{1}{2}}\right\}}\right]$ where V_1 is the escape velocity.

Solution Since V_1 is escape velocity we have from Eq. (9.41)

$$V_1^2 = \frac{2\mu}{r_0}. \qquad \text{(i)}$$

For a rectangular hyperbola $e = \sqrt{2}$, and hence from Eq. (9.42), we get

$$v_0^2 = \mu\left(\frac{2}{r_0} + \frac{1}{L}\right) = V_1^2 + \frac{\mu}{L}$$

or

$$v_0^2 - V_1^2 = \frac{\mu}{L} = \frac{\mu^2}{h^2}, \text{ since } L = \frac{h^2}{\mu}$$

or

$$p_0^2\, v_0^2 (v_0^2 - V_1^2) = \mu^2 \text{ (since } p_0 v_0 = h). \qquad \text{(ii)}$$

Next, using the well known result $p_0 = r_0 \sin\alpha$, wherein α, being the angle between the radius vector and tangent, is the angle of projection, we obtain from Eq. (ii) the required result.

EXAMPLE 9.6 A particle is moving in an elliptical path under the inverse square law of attraction so that apsidal distances are 4 and 5 units. Its speed at apogee is 16 units/s. Find the orbit, and the speed at the perigee.

If the speed of the particle at the perigee is suddenly increased by 50 per cent, find the new orbit. Find the direction which the new orbit approaches and also the ultimate speed.

Solution We are given

$$r_1 = \frac{L}{1+e} = \text{Distance of perigee} = 4 \tag{i}$$

$$r_2 = \frac{L}{1-e} = \text{Distance of apogee} = 5 \tag{ii}$$

Solving Eqs. (i) and (ii), we obtain,

$$L = \frac{40}{9}, \quad e = \frac{1}{9}.$$

Hence, the orbit is

$$\frac{40}{r} = 9 + \cos\theta.$$

Also, we have

$$h = p_2 v_2 = r_2 v_2 = 5.16 = 80,$$

and so get speed at perigee,

$$v_1 = \frac{h}{r_1} = 20 \text{ units/s}$$

also
$$\mu = \frac{h^2}{L} = 1440.$$

Next, representing the quantities by accent when the speed of the particle is increased at perigee, we have

$$v_1' = 20 + 10 = 30$$

$$h' = 4.30 = 120.$$

Since the central force remains the same, the value of μ remains unaltered, and hence, we have

$$L' = \frac{\mu}{h'^2} = \frac{1440}{120 \times 120} = 10. \tag{iii}$$

Further, we have

$$r_1' = r_1 = 4 = \frac{L'}{1+e'},$$

and so using Eq. (iii), we get

$$e' = \frac{3}{2}.$$

Thus, the new orbit is

$$20 = r\,(2 + 3\cos\theta), \tag{iv}$$

Since $e' > 1$, the path is hyperbolic and the direction approached by the orbit is along the asymptote. Thus, in taking the limit $r \to \infty$, we have from Eq. (iv)

$$\cos\theta = -\frac{2}{3},$$

providing the direction of approach.

Also, taking the limit $r \to \infty$, in Eq. (9.42), we get the ultimate speed given by

$$v'^2 = \mu\frac{(e'^2 - 1)}{L'} = \frac{1440.9}{10.4}$$

or

$$v' = 6\sqrt{5} \text{ units/s.}$$

EXAMPLE 9.7 A planet is describing an ellipse about the sun as focus; show that its velocity away from the sun is greatest when the radius vector to the planet is at right angles to the major axis of the path, and it then is $\frac{2\pi ae}{T\sqrt{1-e^2}}$, where $2a$ is the major axis, e the eccentricity, and T the periodic time.

Solution The path is

$$\frac{L}{r} = 1 + e\cos\theta, \tag{i}$$

and so

$$\tan\phi = r\frac{d\theta}{dr} = \frac{L}{re\sin\theta}. \tag{ii}$$

Now, the velocity away from the sun is

$$V = v\cos\phi = \frac{h}{p}\cos\phi \tag{iii}$$

$$= \frac{\sqrt{\mu L}}{r}\cot\phi$$

$$= \sqrt{\frac{\mu}{L}}\, e\sin\theta, \quad \text{[using Eq. (ii)].}$$

Thus, we find that maximum value of V occurs when $\theta = \frac{\pi}{2}$, i.e. when the radius vector to the planet is at right angles to the major axis.

Further, we have from Eq. (iii)

$$V_{\max} = \sqrt{\frac{\mu}{L}}\, e$$

$$= \frac{2\pi ae}{T\sqrt{(1-e^2)}}$$

using Eq. (9.51) and the formula $L = a(1 - e^2)$.

EXAMPLE 9.8 A satellite is projected from the surface of the earth with speed U and that the direction of projection makes an angle α with the upward vertical. Find the value of e and derive a condition for the orbit to be elliptical. Further, find the position of the apses. What instantaneous charge in speed is appropriate at the apogee to put the satellite into a circular orbit? What is the radius of the circular orbit?

Solution We may exploit Eqs. (9.77) and (9.78) to solve the problem. We have here

$$r_0 = R,\ \beta_0 = \frac{\pi}{2} - \alpha,\ v_0 = U,\ \mu = gR^2.$$

Hence, from Eq. (9.77), we get

$$e = \left[1 - \frac{U^2 \sin^2 \alpha}{g^2 R^2}(2gR - U^2)\right]^{\frac{1}{2}}. \qquad \text{(i)}$$

For elliptical orbit $e < 1$, therefore above expression shows that the required condition is $U^2 < 2gR$.

Now, observe that reversing the initial line and then turning it through an angle θ_0, the equation of the ellipse $\frac{L}{r} = 1 + e\cos\theta$ becomes

$$\frac{L}{r} = 1 - e\cos(\theta - \theta_0). \qquad \text{(ii)}$$

Thus, the position of the apses are at $\left(\frac{L}{1-e}, \theta_0\right)$ (apogee) and $\left(\frac{L}{1+e}, \pi + \theta_0\right)$ (perigee)

where
$$L = \frac{h^2}{\mu} = \frac{(pU)^2}{gR^2} = \frac{U^2 \sin^2 \alpha}{g}, \qquad \text{(iii)}$$

and
$$\tan\theta_0 = \frac{\sin\alpha\cos\alpha}{\frac{gR}{U^2} - \sin^2\alpha}. \qquad \text{(iv)}$$

We get Eq. (iv) from Eq. (9.78) by replacing θ by $\pi - \theta_0$.

By imparting additional energy through an instantaneous change in velocity at the apogee the satellite may be put in a circular orbit of radius $\frac{L}{(1-e)}$. Thus, for a circular orbit the speed is given by

$$v = \sqrt{\frac{\mu(1-e)}{L}} = \sqrt{\frac{(1-e)gR^2}{L}}$$

while for original orbit the speed at the apogee is

$$v_0 = \sqrt{\left[\mu\left\{\frac{2}{L}(1-e) - \frac{1-e^2}{L}\right\}\right]} = (1-e)\sqrt{\frac{gR^2}{L}}.$$

Thus, the instantaneous increase in speed is

$$v - v_0 = \left(\sqrt{1-e} - 1 + e\right)\sqrt{\frac{gR^2}{L}}.$$

PROBLEMS

1. A particle is moving under a central force towards the origin, equal to r per unit mass. Prove that

$$\ddot{r} = \frac{h^2}{r^2} - r, \quad \text{where } h \text{ is constant}$$

$$\dot{r}^2 = C - \frac{h^2}{r^2} - r^2, \quad \text{where } C \text{ is constant.}$$

If $h = 6$ and $C = 13$ find the distances of apses from the centre of force and the speeds of the particle at these points. If $C = 0$, what is the nature of the motion.

[**Ans:** $r = 3, 2$ with speeds 2, 3; $C = 0$ implies no movement]

2. A particle is moving under a central force along the curve $r = Ce^{-2\theta}$. Show that the force is proportional to $\frac{1}{r^3}$.

3. O is a fixed point and a particle is projected from a point A at right angles to the line OA with velocity V. The particle is attracted to O with a force $\mu u^3(1 + u^2)$. If OA = 2 units and $V^2 = \frac{9\mu}{32}$, show that the orbit is $r = 2\cos\frac{1}{3}\theta$.

4. A hyperbola has two branches. Show that in scattering under inverse square law, one branch corresponds to an attractive interaction while the other corresponds to a repulsive interaction.

5. A particle describes the following curves under an attractive force $f(r)$ to the pole. Find the central force under which the following orbits are being described.

   (i) Leminscate of Bernoulli;

   (ii) Circle, pole on the circumference

   (iii) $(au)^2 = \frac{\cosh 2\theta - 1}{\cosh 2\theta + 2}$ or $\frac{\cosh 2\theta + 1}{\cosh 2\theta - 2}$

   (iv) $r^n = a^n \cos n\theta$

[**Ans:** (i) $\frac{1}{r^7}$, (ii) $\frac{1}{r^5}$, (iii) $\frac{1}{r^7}$, (iv) $\frac{1}{r^{2n+3}}$]

6. Find the central force under which following orbits are being described:

(i) Ellipse: $\dfrac{b^2}{p^2} = \left(\dfrac{2a}{r}\right) - 1$

(ii) Parabola: $p^2 = \dfrac{1}{2} lr$

(iii) Hyperbola (Near branch): $\dfrac{b^2}{p^2} = \left(\dfrac{2a}{r}\right) + 1$

(iv) Hyperbola (Far branch): $\dfrac{b^2}{p^2} = 1 - \left(\dfrac{2a}{r}\right)$

[**Ans:** Inverse square law; attractive in (i), (ii), (iii) and repulsive in (iv)]

7. The velocity at any point of a central orbit is ($1/n$)th of what it would be for a circular orbit at the same distance; show that the central force varies as $1/r^{2n^2+1}$ and that the equation to the orbit is

$$r^{n^2-1} = a^{n^2-1} \cos[(n^2 - 1)\theta].$$

8. A particle moves under a central force $m\mu[3au^4 - 2(a^2 - b^2)u^5], a > b$; and is projected from an apse at a distance $a + b$ with velocity $\dfrac{\sqrt{\mu}}{(a+b)}$ show that the orbit is $r = a + b \cos \theta$.

9. Obtain the polar equation of the orbit for the case $f(r) = \dfrac{\mu}{r^2}$, and show that in this case, if a particle is projected from a great distance with speed v_0 in a line whose perpendicular distance from O is p, then the closest approach of the particle to O is

$$\frac{\left[\mu + (\mu^2 + p^2 v_0^{\,4})^{\frac{1}{2}}\right]}{v_0^{\,2}}.$$

10. Show that for a central potential of the form $V(r) = -\dfrac{\mu}{r}$ the path will be a circle of radius L when the energy has the value $E = -\dfrac{\mu}{2L}$.

11. A particle moves with a central acceleration. $\dfrac{\mu}{r^2} - \dfrac{\lambda}{r^3}$; show that the apsidal angle is $\dfrac{\pi}{\sqrt{\left(1 + \dfrac{\lambda}{h^2}\right)}}$, where $\dfrac{h}{2}$ is the constant areal velocity.

12. A particle of mass m moves under a central force $m\left[\frac{9}{r^4} - \frac{10}{r^5}\right]$ and is projected from an apse, distance 5 units, with speed $\frac{1}{5}$ unit/s. Show that the orbit can be written in the from $r = 3 + 2\cos\theta$.

13. A particle is describing an ellipse of eccentricity $\frac{1}{2}$. It is desired to change its velocity at apogee so as to make the orbit circular. What charge of speed is required?

$\left[\textbf{Ans:}\ \text{Reduced by factor } \sqrt{\frac{2}{3}}\right]$

14. A particle is projected from the earth's surface with velocity v; show that, if the resistance of the air is neglected, the path is an ellipse of major axis $\frac{2ga^2}{2ga - v^2}$, where a is the radius of the Earth.

15. Show that an unresisted particle falling to the earth's surface from a great distance would acquire a velocity $\sqrt{2ga}$, where a is the earth's radius. Prove that the velocity acquired by a particle similarly falling into the sun is to earth's velocity as the square root of the ratio of the diameter of the earth's orbit to the radius of the sun.

16. If a planet were suddenly stopped in its orbit, supposed circular, show that it would fall into the sun in a time which is $\frac{\sqrt{2}}{8}$ times the period of the planets revolution.

17. The mean distance of Mars from the Sun being 1.524 times that of the Earth, find the time of revolution of Mars about the Sun.

18. A comet describes a parabolic path about the Sun and its perihelion distance is one half the radius of the Earth's orbit, which is assumed to be circular. Prove that the comet remains within the Earth's orbit for $\frac{2}{3\pi}$ of a year.

19. The greatest and least velocities of a certain planet in its orbit round the Sun are 30 km/s and 29.2 km/s, respectively. Find the eccentricity of the orbit.

$\left[\textbf{Ans:}\ \frac{1}{74}\right]$

20. For how many days in the year is the Earth's distance from the Sun greater than the mean distance? What is the ratio of the maximum to the minimum speed of the Earth in its orbit?

[**Ans:** 184.5; 1.034]

21. A comet has speed 5×10^4 m/s when at a distance 2×10^{11} m from the Sun (mass 2×10^{30} kg). Will the orbit be elliptic, parabolic, or hyperbolic?

[**Ans:** Hyperbolic]

22. The acceleration on a satellite is represented by the vector $3\hat{\mathbf{i}} + \hat{\mathbf{j}} + \hat{\mathbf{k}}$. If the position of the satellite relative to the earth is (7, –5, 4), calculate the rate of change of angular momentum relative to the earth given that mass of the satellite is 12 units.

$\left[\textbf{Ans:} \quad 12(-9\hat{\mathbf{i}} + 5\hat{\mathbf{j}} + 22\hat{\mathbf{k}})\right]$

23. Halley's comet when seen in 1910 was moving in a very eccentric orbit with $e = 0.9674$, its perihelion being a distance of 8.77×10^{10} m from the Sun. Halley was born in 1656; how old was he when he observed the comet.

[**Ans:** 26 years]

24. At the end of a launching process a satellite is travelling with velocity v m/s, parallel to the earth's surface and at a height of z metres. Show that the eccentricity of its orbit is given by

$$1 + e = \frac{(R+z)v^2}{Gm}$$

where m and R are the mass and radius of the earth, respectively. Find an expression for the greatest height of the satellite.

$$\left[\textbf{Ans:} \quad \frac{(z+R)^2 v^2}{[2GM - (z+R)v^2]}\right]$$

25. A comet is observed at a distance 1.32×10^{11} m from the Sun, travelling away with a speed of 4.5×10^4 m/s at an angle of 30° to the radius from the Sun. Will the comet even again approach the Sun?

[**Ans:** No]

26. If a communication satellite is to remain in circular orbit constantly above a particular city on the equator, what distance above this city will this orbit be? Why must the city be on the equator?

27. A satellite is launched horizontally at a distance d_0 from the centre of the earth, radius $R < d_0$, with velocity v_0. Show that the distance d and velocity v at another point at which it moves at right angles to the line joining it to the centre are given by $d = \dfrac{d_0}{k}$ and $v = kv_0$ where $k = \left(\dfrac{2gR^2}{d_0 v_0}\right) - 1$. Can there be more than two points at which the satellite moves at right angles to the line joining it to the centre of the earth?

28. A rocket is orbiting the earth in a circular path of altitude H. If the jet engine is activated to give the rocket a sudden burst of speed, derive the expression for the velocity V which must be attained in order to escape from the influence of the earth.

$$\left[\textbf{Ans: } V = R\sqrt{\frac{2g}{(R+H)}},\ R \text{ being Earth's radius}\right]$$

29. If an artificial satellite has a velocity V at perigee at a height H above the earth, determine the radius of curvature of its path at apogee.

30. The eccentricity of a prescribed flight path for a vehicle designed to explore space is 3.0. Determine the initial velocity on escape and the radius vector on escape.

31. A satellite is launched in such a way that when the rocket burns out, its speed is $\frac{1}{4}\sqrt{21\mu}$ units and its distance from the centre of the earth is $\frac{8}{9}$ units. If its direction at this instant makes an angle $\sin^{-1}\left(\frac{9}{2\sqrt{21}}\right)$ with the line joining the centre of the earth O to the satellite P, and the force of attraction is $\frac{m}{OP^2}$ per unit mass, show that the orbit is an ellipse. Find the distances of apogee and perigee and the maximum and minimum speeds of the satellite.

$$\left[\textbf{Ans: } \text{apogee } \frac{4}{3};\ \text{perigee } \frac{4}{5};\ \text{max. speed } \frac{5\sqrt{u}}{4};\ \text{min. speed } \frac{3\sqrt{\mu}}{4}\right]$$

32. Two satellites are moving in circular orbits around a planet. Their altitudes and their periods of revolution are H_1, H_2 and T_1, T_2, respectively. Find an expression for the radius of the planet and show that the acceleration due to gravity at the surface of planet is

$$\frac{4\pi^2(H_2 - H_1)^3}{\left(T_2^{\frac{2}{3}} - T_1^{\frac{2}{3}}\right)\left(H_2 T_1^{\frac{2}{3}} - H_1 T_2^{\frac{2}{3}}\right)^2}$$

(Ignore the gravitational force that one satellite exerts on the other).

10

Motion of a Particle in Three Dimensions

10.1 INTRODUCTION

Let **r** represent the position of a particle, at a time *t*, relative to a fixed origin O. The velocity **v** of the particle is given by the time derivative

$$\mathbf{v} = \frac{d\mathbf{r}}{dt} = \dot{\mathbf{r}}. \tag{10.1}$$

Newton's second law, involving the acceleration **f**, governs the equation of motion. Now

$$\mathbf{f} = \frac{d\mathbf{v}}{dt} = \dot{\mathbf{v}} = \ddot{\mathbf{r}}. \tag{10.2}$$

In Cartesian co-ordinate system (x, y, z), the unit base vectors $\hat{\mathbf{i}}, \hat{\mathbf{j}}, \hat{\mathbf{k}}$ are fixed. Therefore, velocity and acceleration components are simply given by $(\dot{x}, \dot{y}, \dot{z})$ and $(\ddot{x}, \ddot{y}, \ddot{z})$, respectively. But in a general curvilinear co-ordinate system the base vectors are not fixed and so the situation is not that simple. We have to ascertain how do the base vectors change. The analysis is best done using tensor methods. When the co-ordinate system is orthogonal, we may employ vector methods too. We shall present the analysis for spherical polar co-ordinates (r, θ, φ) and cylindrical polar co-ordinates (ρ, φ, z) having unit base vectors $(\mathbf{e}_r, \mathbf{e}_\theta, \mathbf{e}_\varphi)$ and $(\mathbf{e}_\rho, \mathbf{e}_\varphi, \mathbf{e}_z)$, respectively. As the particle changes its position with time, we need the time rate of change of these base vectors determined as follows:

10.2 SPHERICAL POLAR CO-ORDINATES

Now, (r, θ, φ) are related to the Cartesian co-ordinates (x, y, z) through the transformation as follows:

$$x = r\cos\theta,\ \ y = r\sin\theta\cos\varphi,\ \ z = r\sin\theta\sin\varphi. \tag{10.3}$$

The corresponding scale factors for (r, θ, φ) are given by

$$h_r = \sqrt{\left(\frac{\partial x}{\partial r}\right)^2 + \left(\frac{\partial y}{\partial r}\right)^2 + \left(\frac{\partial z}{\partial r}\right)^2} = 1, \tag{10.4}$$

$$h_\theta = \sqrt{\left(\frac{\partial x}{\partial \theta}\right)^2 + \left(\frac{\partial y}{\partial \theta}\right)^2 + \left(\frac{\partial z}{\partial \theta}\right)^2} = r, \tag{10.4a}$$

$$h_\varphi = \sqrt{\left(\frac{\partial x}{\partial \varphi}\right)^2 + \left(\frac{\partial y}{\partial \varphi}\right)^2 + \left(\frac{\partial z}{\partial \varphi}\right)^2} = r\sin\theta. \tag{10.4b}$$

Noting that the direction cosines of the $\mathbf{e}_r$ direction are given by

$$\left(\frac{1}{h_r}\frac{\partial x}{\partial r}, \frac{1}{h_r}\frac{\partial y}{\partial r}, \frac{1}{h_r}\frac{\partial z}{\partial r}\right) \equiv (\cos\theta, \sin\theta\cos\varphi, \sin\theta\sin\varphi)$$

and similar expressions for $\mathbf{e}_\theta$, $\mathbf{e}_\varphi$, we can draw the following for transformation from $\hat{\mathbf{i}}, \hat{\mathbf{j}}, \hat{\mathbf{k}}$ to $\mathbf{e}_r$, $\mathbf{e}_\theta$, $\mathbf{e}_\varphi$ and vice-versa

$$\left.\begin{array}{cccc} & \hat{\mathbf{i}} & \hat{\mathbf{j}} & \hat{\mathbf{k}} \\ \mathbf{e}_r & \cos\theta & \sin\theta\cos\varphi & \sin\theta\sin\varphi \\ \mathbf{e}_\theta & -\sin\theta & \cos\theta\cos\varphi & \cos\theta\sin\varphi \\ \mathbf{e}_\varphi & 0 & -\sin\theta\sin\varphi & \sin\theta\cos\varphi \end{array}\right\} \tag{10.5}$$

Remark: The transformation in Expression (10.5) may also be written by geometrically visualizing the directions of base vectors $\mathbf{e}_r$, $\mathbf{e}_\theta$, $\mathbf{e}_\varphi$.

Taking the time derivatives, we now get from Expression (10.5) the time rates of change of base vectors $\mathbf{e}_r$, $\mathbf{e}_\theta$, $\mathbf{e}_\varphi$ as

$$\left.\begin{array}{cccc} & \hat{\mathbf{i}} & \hat{\mathbf{j}} & \hat{\mathbf{k}} \\ \dot{\mathbf{e}}_r & -\sin\theta\dot{\theta} & \cos\theta\cos\varphi\dot{\theta} - \sin\theta\sin\varphi\dot{\varphi} & \cos\theta\sin\varphi\dot{\theta} + \sin\theta\cos\varphi\dot{\varphi} \\ \dot{\mathbf{e}}_\theta & -\cos\theta\dot{\theta} & -\sin\theta\cos\varphi\dot{\theta} - \cos\theta\sin\varphi\dot{\varphi} & -\sin\theta\sin\varphi\dot{\theta} + \cos\theta\cos\varphi\dot{\varphi} \\ \dot{\mathbf{e}}_\varphi & 0 & -\cos\theta\sin\varphi\dot{\theta} - \sin\theta\cos\varphi\dot{\varphi} & \cos\theta\sin\varphi\dot{\theta} - \sin\theta\sin\varphi\dot{\varphi} \end{array}\right\} \tag{10.6}$$

Since $\mathbf{e}_r$ is a vector of constant magnitude (hence $\dot{\mathbf{e}}_r$ is in the plane of $\mathbf{e}_\theta$ and $\mathbf{e}_\varphi$), we can write $\dot{\mathbf{e}}_r$ = A $\mathbf{e}_\theta$ + B $\mathbf{e}_\varphi$, whence taking dot product with $\mathbf{e}_\theta$ and $\mathbf{e}_\varphi$, respectively we get A = $\mathbf{e}_\theta \cdot \dot{\mathbf{e}}_r$, and B = $\mathbf{e}_\varphi \cdot \dot{\mathbf{e}}_r$. Next, using Expressions (10.5) and (10.6), we can easily obtain the values of A and B, and hence of $\dot{\mathbf{e}}_r$. Similarly, we can evaluate the values of $\dot{\mathbf{e}}_\theta$ and $\dot{\mathbf{e}}_\varphi$. Collecting these quantities, we have

$$\left.\begin{aligned} \dot{\mathbf{e}}_r &= \dot{\theta}\,\mathbf{e}_\theta + \sin\theta\,\dot{\varphi}\,\mathbf{e}_\varphi \\ \dot{\mathbf{e}}_\theta &= -\,\dot{\theta}\,\mathbf{e}_r + \cos\theta\,\dot{\varphi}\,\mathbf{e}_\varphi \\ \dot{\mathbf{e}}_\varphi &= -\sin\theta\,\dot{\theta}\,\mathbf{e}_r - \cos\theta\,\dot{\varphi}\,\mathbf{e}_\theta \end{aligned}\right\} \qquad (10.7)$$

Now, using first of Eq. (10.7), the velocity $\mathbf{v}$ of the particle is given by

$$\begin{aligned} \mathbf{v} &= \frac{d\mathbf{r}}{dt} = \dot{r}\mathbf{e}_r + r\dot{\mathbf{e}}_r \\ &= \dot{r}\mathbf{e}_r + r\dot{\theta}\,\mathbf{e}_\theta + r\sin\theta\,\dot{\varphi}\,\mathbf{e}_\varphi. \end{aligned} \qquad (10.8)$$

Again, the acceleration $\mathbf{f}$ may be obtained by taking the time derivative of $\mathbf{v}$ and utilizing Eq. (10.7). Thus, after simplification, we get

$$\begin{aligned} \mathbf{f} &= (\ddot{r} - r\dot{\theta}^2 - r\sin^2\theta\dot{\varphi}^2)\,\mathbf{e}_r + (r\ddot{\theta} + 2\dot{r}\dot{\theta} - r\sin\theta\cos\theta\dot{\varphi}^2)\,\mathbf{e}_\theta \\ &\quad + (r\sin\theta\,\ddot{\varphi} + 2\sin\theta\,\dot{r}\dot{\varphi} + 2r\cos\theta\,\dot{\theta}\,\dot{\varphi})\,\mathbf{e}_\varphi. \end{aligned} \qquad (10.9)$$

10.3 CYLINDRICAL POLAR CO-ORDINATES

We can calculate the expressions for velocity $\mathbf{v}$ and acceleration $\mathbf{f}$ in cylindrical polar co-ordinates ρ, φ, z

$$\mathbf{v} = \dot{\rho}\,\mathbf{e}_\rho + \rho\,\dot{\varphi}\,\mathbf{e}_\varphi + \dot{z}\,\mathbf{e}_z, \qquad (10.10)$$

$$\mathbf{f} = (\ddot{\rho} - \rho\,\dot{\varphi}^2)\,\mathbf{e}_\rho + (\rho\,\ddot{\varphi} + 2\dot{\rho}\dot{\varphi})\,\mathbf{e}_\varphi + \ddot{z}\,\mathbf{e}_z. \qquad (10.11)$$

We can easily arrive at Eqs. (10.10) and (10.11) for the three dimensional cylindrical polar components from the uniplanar results (obtained in Chapter 7) by replacing r by ρ and by θ by φ and adding the axial velocity component $\dot{z}$ and the axial acceleration component $\ddot{z}$ to those results respectively.

Equations (10.8) and (10.9) for velocity and acceleration in spherical polar co-ordinates may be obtained from Eqs. (10.10) and (10.11) in cylindrical polar co-ordinates by making use of the transformations between the two systems as follows:

$$\rho = r\sin\theta,\ z = r\cos\theta,\ \varphi = \varphi. \qquad (10.12)$$

$$\mathbf{e}_\rho = \sin\theta\,\mathbf{e}_r + \cos\theta\,\mathbf{e}_\theta,\ \mathbf{e}_z = \cos\theta\,\mathbf{e}_r - \sin\theta\,\mathbf{e}_\theta, \qquad (10.13)$$

$$\ddot{\rho}\mathbf{e}_\rho + \ddot{z}.\mathbf{e}_z = (\ddot{r} - r\dot{\theta}^2)\,\mathbf{e}_r + (r\ddot{\theta} + 2\dot{r}\dot{\theta})\,\mathbf{e}_\theta + \ddot{z}\mathbf{e}_z. \qquad (10.14)$$

Equation (10.14) has been written down by observing that in the $\rho - z$ plane (r, θ) may be considered as plane polar co-ordinates.

10.4 KINEMATICS IN NATURAL CO-ORDINATES

Let s be the arc distance of a particle moving on a curve, and $\mathbf{r}$ be its field position at time t. Further, let $\mathbf{t}$, $\mathbf{n}$, $\mathbf{b}$, representing unit tangent, unit principal normal and unit binormal respectively form the natural trihedron associated with the trajectory. From our knowledge of geometry, we know that

$$\frac{d\mathbf{r}}{dt} = \mathbf{t}, \frac{d\mathbf{t}}{ds} = \kappa.\mathbf{n}, \tag{10.15}$$

where κ is the curvature of the path curve.

Now, the velocity of the particle is given by

$$\mathbf{v} = \frac{d\mathbf{r}}{dt} = \frac{ds}{dt}\frac{d\mathbf{r}}{ds} = v\mathbf{t}, \tag{10.16}$$

where $v = \dfrac{ds}{dt}$ is the speed of the particle.

Equation (10.16) shows that the particle has only the tangential component of velocity.

Next, the acceleration is given by

$$\mathbf{f} = \frac{d\mathbf{v}}{dt} = \frac{d(v\mathbf{t})}{dt} = \frac{dv}{dt}\mathbf{t} + v\frac{ds}{dt}\frac{d\mathbf{t}}{ds} = v\frac{dv}{ds}\mathbf{t} + \kappa v^2\,\mathbf{n}. \tag{10.17}$$

Thus, we see that acceleration has only two components, $\dfrac{dv}{dt} = \dfrac{vdv}{ds}$ in the tangential direction and κv^2 in the direction of the principal normal, and there is no component in the direction of the binormal i.e., the motion is in the osculating plane at the point.

It may be noted that from Euler's Theorem in geometry,

$$\kappa = \kappa_1 \sin^2 \lambda + \kappa_2 \cos^2 \lambda,$$

where κ_1 and κ_2 denote the greatest and least curvatures of normal sections and λ is the angle between the direction of motion and the direction of maximum curvature.

10.5 ROTATING FRAME OF REFERENCE

An important example of non-inertial frame is that of frame rotating about its axis. It will be seen that fictitious forces like centrifugal force and Coriolis force appear in the case of a frame fixed with respect to rotating earth. These forces are negligible in most situations but for long range events such as flight of guided missiles, rockets and winds Coriolis force becomes important. We shall first derive the expressions for velocity and acceleration.

Let O be the common origin of two co-ordinate systems S and S′ having z-axis as a common axis. Let S be a fixed co-ordinate system and S′ rotate about the common z-axis with angular velocity $\boldsymbol{\omega} = \omega \mathbf{e}_z$, where ω is in general, time dependent but $\mathbf{e}_z$ is constant unit vector along the common z-axis which is the axis of rotation of frame S′ with respect to the fixed frame S. Since S or S′ have always the same pole O, a point P in space will have the same position vector **r** (= **OP**) in either of the frame S or S′. In the cylindrical polar co-ordinate system (ρ, φ, z) fixed in S′, we can write

$$\mathbf{r} = \rho\, \mathbf{e}_\rho + z\, \mathbf{e}_z. \tag{10.18}$$

Suppose at some instance of time (say $t = 0$) the S and S′ are aligned, and then in frame S also Eq. (10.10) holds. But while $\mathbf{e}_z$ is fixed, as frame S′ rotates the unit vectors $\mathbf{e}_\rho$ and $\mathbf{e}_\varphi$ will also rotate and we have

$$\dot{\mathbf{e}}_\rho = \omega\, \mathbf{e}_\varphi \text{ and } \dot{\mathbf{e}}_\varphi = -\,\omega\, \mathbf{e}_\rho. \tag{10.19}$$

Now, consider the motion of point P in space. We shall represent the time derivatives in S and S′ respectively by $\frac{\delta}{\delta t}$ and $\frac{d}{dt}$. Thus, we have from Eq. (10.10)

$$\mathbf{u} = \frac{\delta \mathbf{r}}{\delta t} = \dot{\rho}\, \mathbf{e}_\rho + \rho \dot{\mathbf{e}}_\rho + \dot{z}\, \mathbf{e}_z = \dot{\rho}\, \mathbf{e}_\rho + \dot{z}\, \mathbf{e}_z + \rho \dot{\mathbf{e}}_\rho \tag{10.20}$$

$$= \dot{\rho}\, \mathbf{e}_\rho + \dot{z}\, \mathbf{e}_z + \rho \omega \dot{\mathbf{e}}_\varphi$$

$$= \frac{d\mathbf{r}}{dt} + \boldsymbol{\omega} \times \mathbf{r} = \mathbf{v} + \boldsymbol{\omega} \times \mathbf{r} \tag{10.21}$$

or

$$\mathbf{v} = \mathbf{u} - \boldsymbol{\omega} \times \mathbf{r}.$$

This provides the relation between the velocity **v** in the rotating system and velocity **u** in the fixed system.

Next, taking $\mathbf{v} = \frac{d\mathbf{r}}{dt} + \boldsymbol{\omega} \times \mathbf{r}$ in place of **r**, and applying the same process as above, we get the acceleration in the fixed system as

$$\mathbf{f} = \frac{d^2 \mathbf{r}}{dt^2} = \frac{d\mathbf{v}}{dt} + \boldsymbol{\omega} \times \mathbf{v} \tag{10.22}$$

$$= \frac{d}{dt}\left[\frac{d\mathbf{r}}{dt} + \boldsymbol{\omega} \times \mathbf{r}\right] + \boldsymbol{\omega} \times \left[\frac{d\mathbf{r}}{dt} + \boldsymbol{\omega} \times \mathbf{r}\right]$$

$$= \frac{d^2 \mathbf{r}}{dt^2} + 2\boldsymbol{\omega} \times \frac{d\mathbf{r}}{dt} + \dot{\boldsymbol{\omega}} \times \mathbf{r} + \boldsymbol{\omega} \times (\boldsymbol{\omega} \times \mathbf{r}).$$

providing acceleration **a** in the rotating system in terms of the acceleration **f** in the fixed system as

$$\frac{d^2 \mathbf{r}}{dt^2} = \mathbf{f} - 2\boldsymbol{\omega} \times \frac{d\mathbf{r}}{dt} - \dot{\boldsymbol{\omega}} \times \mathbf{r} - \boldsymbol{\omega} \times (\boldsymbol{\omega} \times \mathbf{r}) \tag{10.23}$$

$$= \mathbf{f} - 2\boldsymbol{\omega} \times \frac{d\mathbf{r}}{dt} - \dot{\boldsymbol{\omega}} \times \mathbf{r} + \{(\boldsymbol{\omega} \cdot \mathbf{r})\boldsymbol{\omega} - (\boldsymbol{\omega} \cdot \boldsymbol{\omega})\mathbf{r}\}.$$

Here the acceleration **f** is same as the body force on a unit mass in the fixed coordinate system which may be taken as fixed with respect to the Sun or with respect to the distance stars. In the rotating co-ordinate system, the first term on the right hand side is the true body force and the other three terms are fictitious forces, arising out of the use of a non-inertial co-ordinate system, The term $-2\boldsymbol{\omega} \times \frac{d\mathbf{r}}{dt} = -2\boldsymbol{\omega} \times \mathbf{v}$ is the *Coriolis force* and the term $-\boldsymbol{\omega} \times (\boldsymbol{\omega} \times \mathbf{r}) = \omega^2 \mathbf{r} - (\boldsymbol{\omega} \cdot \mathbf{r})\boldsymbol{\omega}$ is the centrifugal force. The small effect of Coriolis force is amplified by focult pendulum to demonstrate the rotation of the earth in laboratory.

Many a times it is convenient to work in suitably chosen Cartesian co-ordinate system. From Eq. (10.23), we can easily write down the component equations in Cartesian system as follows:

$$\left.\begin{aligned} \ddot{x} &= f_x - 2(\omega_y \dot{z} - \omega_z \dot{y}) - (\dot{\omega}_y z - \omega_z \dot{y}) + \{\omega^2 x - \omega_x (\omega_x x + \omega_y y + \omega_z)\} \\ \ddot{y} &= f_y - 2(\omega_z \dot{x} - \omega_x \dot{z}) - (\dot{\omega}_z x - \dot{\omega}_x z) + \{\omega^2 y - \omega_y (\omega_x x + \omega_y y + \omega_z)\} \\ \ddot{z} &= f_z - 2(\omega_x \dot{y} - \omega_y \dot{x}) - (\dot{\omega}_x y - \dot{\omega}_y x) + \{\omega^2 z - \omega_z (\omega_x x + \omega_y y + \omega_z)\} \end{aligned}\right\} \tag{10.24}$$

where dots represent differentiation with respect to time and $\omega^2 = \omega_x^2 + \omega_y^2 + \omega_z^2$.

10.6 THE ROTATING EARTH

Let us consider a co-ordinate system attached to the earth, rotating with an uniform angular velocity $\boldsymbol{\omega}$ ($\dot{\boldsymbol{\omega}} = 0$) as our rotating co-ordinate system. Further, since the angular speed $\omega \left(\simeq \frac{2\pi}{158400} \text{ radians per second} \right)$ is a small quantity, the centrifugal force being $O(\omega^2)$ is negligible. The Coriolis force $-2\boldsymbol{\omega}\times\mathbf{v}$, although small for laboratory observations, may become appreciable when **v** is substantial or when the time duration of the event is large as in the case of cyclones and missiles motion. Thus, for a particle of unit mass in the rotating earth co-ordinate system, we have the following simplified equation of motion

$$\frac{d^2 \mathbf{r}}{dt^2} = \mathbf{f} - 2\boldsymbol{\omega} \times \mathbf{v}, \text{ where } \mathbf{v} = \frac{d\mathbf{r}}{dt}. \tag{10.25}$$

We shall take a co-ordinate system with origin O at a point on the surface of the earth (taken tc be a sphere) and Cartesian axis x, y, z attached to it. It is found convenient to take one of the axes (say y-axis) along the vertical there and x and z axes along suitably chosen mutually perpendicular tangential directions (to the sphere). In this way the Cartesian equations of motion follow from Eq. (10.25) as

$$\left.\begin{aligned} \ddot{x} &= -2(\omega_y\dot{z} - \omega_z\dot{y}) - (\dot{\omega}_y z - \dot{\omega}_z y) \\ \ddot{y} &= f_v - 2(\omega_z\dot{x} - \omega_x\dot{z}) - (\dot{\omega}_z x - \dot{\omega}_x z) \\ \ddot{z} &= -2(\omega_x\dot{y} - \omega_y\dot{x}) - (\dot{\omega}_x y - \dot{\omega}_y x) \end{aligned}\right\} \tag{10.26}$$

Here f_y is the gravitational force given by Newton's gravitational force, and near the earth's surface has the value $-g$. It should be kept in view that the $O(\omega^2)$ terms have been neglected while deriving Eq. (10.26).

SOLVED EXAMPLES

EXAMPLE 10.1 A smooth circular cone, of angle 2α, has its axis vertical and its vertex, which is pierced with a small hole, downwards. A mass M hangs at rest by a string which passes through the vertex, and a mass m is attached to the upper end and describes a horizontal circle on the inner surface of the cone. Find the time T of a complete revolution, and show that small oscillations about the steady motion takes place in time $T \operatorname{cosec} \alpha \sqrt{\dfrac{(M+m)}{3m}}$.

Solution The masses m and M are attached to the ends of a string of length L. Taking the vertex of the cone as the pole we set up spherical polar co-ordinate system (R, θ, φ). Here the polar angle $\theta = \alpha$ is constant and hence $\dot{\theta} = 0$.

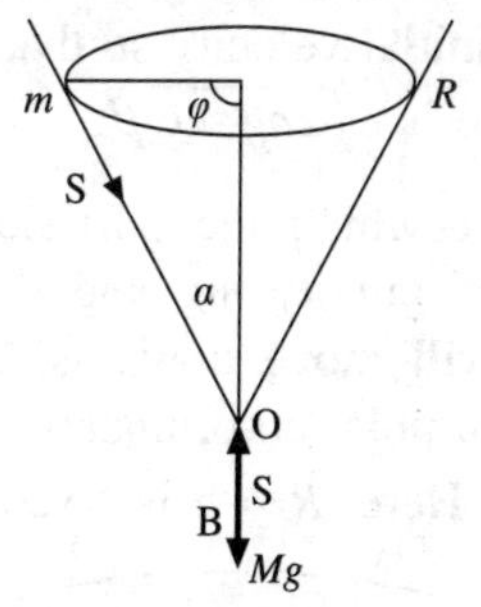

Figure 10.1 Masses m and M attached to the ends A and B of the string passing through a hole at O.

With the force system as depicted in Figure 10.1, the polar and azimuthal equations of motion for the mass m may be expressed as

$$m(\ddot{R} - R\sin^2\alpha\,\dot{\varphi}^2) = -S \tag{i}$$

$$R\ddot{\varphi} + 2\dot{R}\dot{\varphi} = 0. \tag{ii}$$

Also, for the motion of mass M, we have

$$-M\ddot{R} = Mg - S. \tag{iii}$$

Eliminating S in between Eq. (i) and (iii), we get

$$(m + M)\,\ddot{R} - mR\sin^2\alpha\,\dot{\varphi}^2 = -Mg. \tag{iv}$$

Here S is the tension in the string. When in steady motion, let m move round a circle of constant radius $R_0 \sin\alpha$. Now, Eq. (ii) implies $R_0^2\dot{\varphi} = k$, a constant, providing angular velocity $\dot{\varphi} = \frac{k}{R_0^2}$. Then the time for a complete revolution is

$$T = \frac{2\pi}{\dot{\varphi}} = \frac{2\pi R_0^2}{k}. \tag{v}$$

For small oscillations about the mean position, we may write $R = R_0 + x$, where x is a small perturbation. Substituting the value of R in Eq. (iv) and neglecting terms of $O(x^2)$ and simplifying, we get the following differential equation for x:

$$\ddot{x} = -\frac{3mk^2\sin^2\alpha}{(m+M)\,R_0^4}\,x + \frac{mk^2\sin^2\alpha}{R_0^3} - Mg. \tag{vi}$$

This may be recognized as simple harmonic equation with time of complete oscillation given by

$$T\,\text{cosec}\,\alpha\sqrt{\frac{M+m}{3m}} \quad \text{(substituting for } k \text{ in terms of } R_0). \tag{vii}$$

EXAMPLE 10.2 A particle is projected horizontally along the interior surface of a smooth hemisphere whose axis is vertical and whose vertex is downwards; the point of projection being at an angular distance β from the lowest point, show that the initial velocity so that the particle may just ascend to the rim of the hemisphere is $\sqrt{2ag\sec\beta}$.

Solution The forces under which the particle of mass m moves on the hemisphere are the force of gravity mg and the reaction S as depicted in Figure 10.2. The particle will move upwards, as reaction S has a vertical component. We use spherical polar co-ordinates (R, θ, φ) with the centre O of the hemisphere as the pole. Here, $R = a$ is fixed and so $\dot{R} = \ddot{R} = 0$.

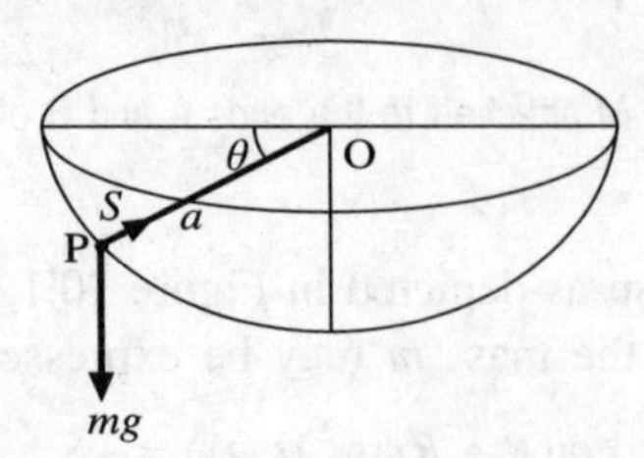

Figure 10.2 Particle P moving in the interior of smooth hemisphere.

The equations of motion in the transverse and azimuthal directions may now be written as

$$a\,(\ddot{\theta} - \sin\theta\cos\theta\dot{\varphi}^2) = g\cos\theta \tag{i}$$

$$a\,(\sin\theta\ddot{\varphi} + 2\cos\theta\dot{\theta}\dot{\varphi}) = 0. \tag{ii}$$

Equation (ii), on integration, yields

$$a\sin^2\theta\dot{\varphi} = \text{Constant} = V\sin\beta, \tag{iii}$$

where the value of the constant has been evaluated by using the initial condition that at $\theta = \beta$ the (horizontal) azimuthal velocity $a\sin\beta\dot{\varphi} = V$. Substituting the value of $\dot{\varphi}$ in Eq. (i), we get

$$a\left(\ddot{\theta} - \frac{V^2\sin^2\beta\cos\theta}{a^2\sin^3\theta}\right) = -\,g\sin\theta. \tag{iv}$$

Multiplying Eq. (iv) by $2\dot{\theta}$ on both sides, and then integrating, we obtain

$$\dot{\theta}^2 = -\frac{V^2\sin^2\beta}{a\sin^2\theta} + 2g\cos\theta + C. \tag{v}$$

Now, we apply the conditions that $\dot{\theta} = 0$ at $\theta = \beta$ and at $\theta = \dfrac{\pi}{2}$ to get

$$0 = -\frac{V^2}{a} + 2g\cos\beta + C, \tag{vi}$$

and

$$0 = -\frac{V^2\sin^2\beta}{a} + C. \tag{vii}$$

Eliminating C between Eqs. (vi) and (vii), we easily get the required value of V.

EXAMPLE 10.3 A particle is constrained to move on a smooth conical surface of vertical angle 2α, and describes a plane curve under the action of an attraction to the vertex, the plane of the orbit cutting the axis of the cone at a distance a from the vertex. Show that the attractive force must vary as $\dfrac{1}{R^2} - \dfrac{a\cos\alpha}{R^3}$.

Solution Let us use spherical polar co-ordinates (R, θ, φ) with the vertex O of the cone as pole and $\mathbf{e}_R$, $\mathbf{e}_\theta$, $\mathbf{e}_\varphi$ as the corresponding base vectors. On the cone $\theta = \alpha$ (Figure 10.3), we have $\dot{\theta} = \ddot{\theta} = 0$. If $f(R)$ is the attractive force, the equations of motion in the polar and azimuthal directions may be written down as

$$\ddot{R} - R\sin^2\alpha\dot{\varphi}^2 = -\,f(R) \tag{i}$$

$$R\ddot{\varphi} + 2\dot{R}\dot{\varphi} = 0. \tag{ii}$$

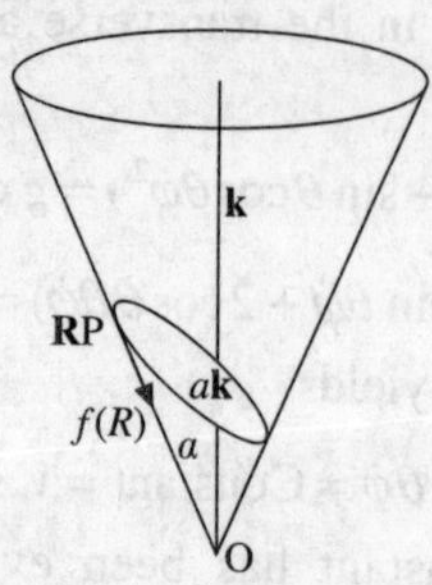

Figure 10.3 Particle P constrained to move on a smooth conical surface under attractive force $f(R)$.

Equation (ii) on integration gives $\dot{\varphi} = \dfrac{h}{R^2}$ where h is a constant. Using the value of $\dot{\varphi}$ in Eq. (i) we get

$$f(R) = \frac{h^2 \sin^2 \alpha}{R^3} - \ddot{R}. \tag{iii}$$

Now, we use the condition that the path is a plane curve to derive the value of $\ddot{R}$ in terms of R to derive the form of the function $f(R)$. If $\mathbf{k}$ is unit vector along the axis of the cone, $a\mathbf{k}$ is the point where the axis of the cone cuts the plane of the curve. Using the fact that the direction of the path is along the direction of the velocity and that $\mathbf{R} - a\mathbf{k}$ is a vector in the plane of the curve, we can express the unit normal to this plane as

$$\mathbf{n} = \frac{(R\mathbf{e}_R - a\mathbf{k}) \times (\dot{R}\mathbf{e}_R + R \sin\alpha\, \dot{\varphi}\, \mathbf{e}_\varphi)}{\text{Modulus of the numerator}} \tag{iv}$$

$$= \frac{aR\dot{\varphi}\mathbf{e}_r - R^2 \dot{\varphi}\mathbf{e}_\theta - aR\dot{R}\,\mathbf{e}_\varphi}{\sqrt{a^2 \dot{R}^2 + R^2 (R^2 + a^2 - 2aR\cos\alpha)\dot{\varphi}^2}},$$

where

$$\mathbf{e}_r = \mathbf{e}_R \sin\alpha + \mathbf{e}_\theta \cos\alpha.$$

Now, if β denotes the angle between the normal to the plane and its axis, we have

$$\mathbf{k} \cdot \mathbf{n} = \frac{R^2 \sin\alpha\, \dot{\varphi}}{\sqrt{a^2 \dot{R}^2 + R^2 (R^2 + a^2 - 2aR\cos\alpha)\dot{\varphi}^2}} = \cos\beta. \tag{v}$$

Note: The students are advised to work out the value of R^2 from Eq. (v) and then differentiate it to get $\ddot{R}$ which, when substituted in Eq. (iii), provides the desired result.

EXAMPLE 10.4 A particle moves on a rough circular cylinder under the action of no external force. Initially, the particle has a velocity V in a direction making an angle α with the transverse plane of the cylinder; show that the space described in time t is $\left(\dfrac{a}{\mu\cos^2\alpha}\right)\log\left[1+\left(\dfrac{\mu V\cos^2\alpha}{a}\right)t\right]$.

Solution The particle moves under the force of normal reaction S (per unit mass) to the surface of the cylinder and the force of friction, acting tangentially to the path as marked in Figure 10.4.

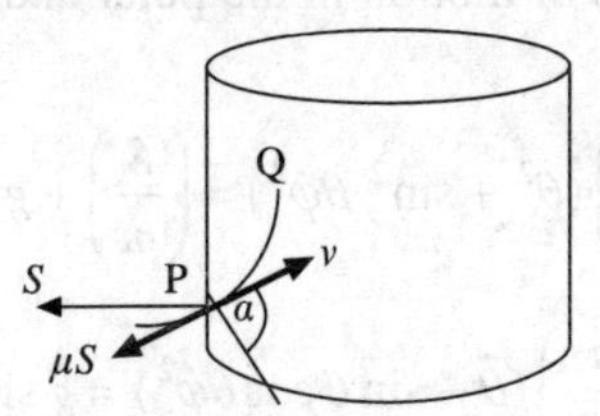

Figure 10.4 Particle P moving on the path PQ on a rough circular cylinder.

The equations of motion in the tangential and axial directions are

$$\frac{dv}{dt}=v\frac{dv}{ds}=-\mu S \tag{i}$$

$$\frac{d^2z}{dt^2}=-\mu S\sin\lambda=\left(\frac{dv}{dt}\right)\left(\frac{dz}{ds}\right)=\left(\frac{1}{v}\frac{dv}{dt}\right)\left(\frac{dz}{dt}\right), \tag{ii}$$

where $\dfrac{dz}{ds}=\sin\lambda$ provides the inclination of the path to the z direction. Integration of Eq. (ii) will show that $\lambda=\text{Constant}=\alpha,$ the initial value. Again, the motion in a direction normal (radial) to the cylinder is given by

$$\frac{v^2\cos^2\alpha}{a}=S. \tag{iii}$$

Eliminating S from Eqs. (i) and (iii), we obtain

$$v\frac{dv}{ds}=-\frac{\mu\cos^2\alpha}{a}v^2, \tag{iv}$$

which on integration yields

$$v=\frac{ds}{dt}=V\exp\left(-\frac{\mu\cos^2\alpha}{a}s\right), \tag{v}$$

where we have set $v = V$ when $s = 0$. Integrating Eq. (v) we get the required result under the initial condition, $s = 0$ at $t = 0$. (The students are advised to work out).

EXAMPLE 10.5 A heavy particle is projected on the inner surface of a smooth spherical shell of radius $\dfrac{a}{\sqrt{2}}$ with velocity $\sqrt{\dfrac{7ag}{3}}$ at a depth $\dfrac{2a}{3}$

below the centre. Show that it will rise to a height $\frac{a}{3}$ above the centre, and that the pressure on the sphere vanishes there.

Solution The forces on the particle of mass m are marked as in Figure 10.5. On the sphere $R = \frac{a}{\sqrt{2}}$, we have $\dot{R} = \ddot{R} = 0$, and hence, with forces as marked in the figure the equations of motion in the polar and transverse directions may be expressed as

$$\left(\frac{a}{\sqrt{2}}\right)(\dot{\theta}^2 + \sin^2\theta\dot{\varphi}^2) = \left(\frac{S}{m}\right) + g\cos\theta, \tag{i}$$

$$\left(\frac{a}{\sqrt{2}}\right)(\ddot{\theta} - \sin\theta\cos\theta\dot{\varphi}^2) = g\sin\theta. \tag{ii}$$

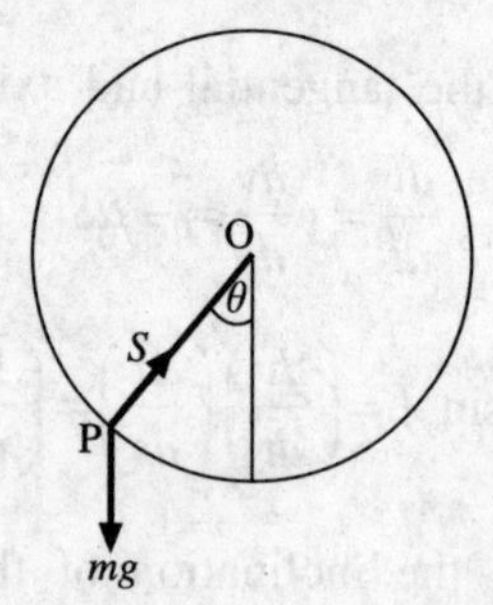

Figure 10.5 Particle P moving on the inner surface of smooth sphere.

Instead of writing the third equation in the azimuthal direction it is more convenient to use the energy equation given as

$$\frac{1}{2}\left(\frac{7ag}{3} - \frac{a^2\dot{\theta}^2}{2} - \frac{a^2\sin^2\theta\dot{\varphi}^2}{2}\right) = \left(\frac{a\cos\theta}{\sqrt{2}} + \frac{2a}{3}\right)g. \tag{iii}$$

Now, substituting the value of $\dot{\varphi}^2$ from Eq. (ii) in Eq. (iii) and simplifying, we obtain the following differential equation for θ:

$$\ddot{\theta}\sin\theta + \dot{\theta}^2\cos\theta = \frac{2g(\sqrt{2} + 2\cos\theta - 3\sqrt{2}\cos^2\theta)}{a}. \tag{iv}$$

Multiplying both sides of the above equation by $2\dot{\theta}\sin\theta$, and integrating, we get

$$(\dot{\theta}\sin\theta)^2 = \frac{2g\left\{\left(\frac{20}{27}\right) - \sqrt{2}\cos\theta - \cos^2\theta + \sqrt{2}\cos^3\theta\right\}}{a}, \tag{v}$$

where the constant of integration has been obtained by the prescribed condition viz., at $t = 0$, $\dot{\theta} = 0$ and $\cos\theta = -\left(\frac{2\sqrt{2}}{3}\right)$. Now the highest point will be reached when $\dot{\theta}$ vanishes again. It may be checked from Eq. (v) that $\dot{\theta} = 0$ again when $\cos\theta = \frac{\sqrt{2}}{3}$, $\left(\sin\theta = \frac{\sqrt{7}}{3}\right)$, and then from Eq. (iii), we have $a\,(\dot{\varphi}^2 \sin^2\theta) = \frac{2g}{3}$. Substituting these values, we have from Eq. (i)

$$\frac{S}{m} = \frac{a}{\sqrt{2}}\dot{\varphi}^2 \sin^2\theta - g\cos\theta = g\left(\frac{\sqrt{2}}{3} - \frac{\sqrt{2}}{3}\right) = 0$$

showing that the pressure vanishes.

EXAMPLE 10.6 Show that the force of gravity apparently increases on account of the rotation of the earth as we move from a point at the pole to a point on the equator.

Solution Let the uniform speed of the earth's rotation be $\boldsymbol{\omega}$ (a constant) and take a point P at a distance r from its centre. Since Coriolis force has no effect on a fixed point and $\boldsymbol{\omega}$ is a constant, the only fictitious force is the centrifugal force

$$\mathbf{f}(r) = -\boldsymbol{\omega}\mathbf{x}(\boldsymbol{\omega}\mathbf{x}\mathbf{r}). \tag{i}$$

Now, the force of gravity is towards the centre of the earth and so may be expressed as

$$-\,\mathrm{F}(r)\,\mathbf{e}_r. \tag{ii}$$

Thus, the increase in the force of gravity due to the rotation of the earth is given by

$$-\mathbf{f}(r).\mathbf{e}_r = \omega^2 r \sin^2\lambda, \tag{iii}$$

where λ is the latitude of the point.

Equation (iii) clearly shows that the apparent gravity increases as we move from the pole ($\lambda = 0$) to the equator $\left(\lambda = \frac{\pi}{2}\right)$.

EXAMPLE 10.7 Let a point O on the earth's surface be at latitude λ in the northern hemisphere. A missile is projected at an angle α from O in the eastward direction with speed V and remains close to O. Neglecting $O(\omega^2)$ terms show that the missile will strike at a point on the horizontal plane through O at a point whose perpendicular distance from the plane of projection is

$$\frac{4\omega V^3 \sin^2\alpha\cos\alpha\sin\lambda}{g^2}$$

towards the south. Also, find the distance travelled along the plane of projection.

Solution Let us take a co-ordinate system with x-axis towards north, y-axis vertically upwards and z-axis towards the east; in this way y-z plane is the plane of projection and z-x the horizontal plane through the point of projection O.

The angle made by the vertical y-axis, passing through the centre of the earth, with the axis of rotation of the earth is $\frac{\pi}{2} - \lambda$ and this axis lies in the x-y plane, hence, it may be seen that angular velocity $\boldsymbol{\omega} = \omega(\mathbf{e}_x \cos\lambda + \mathbf{e}_y \sin\lambda)$.

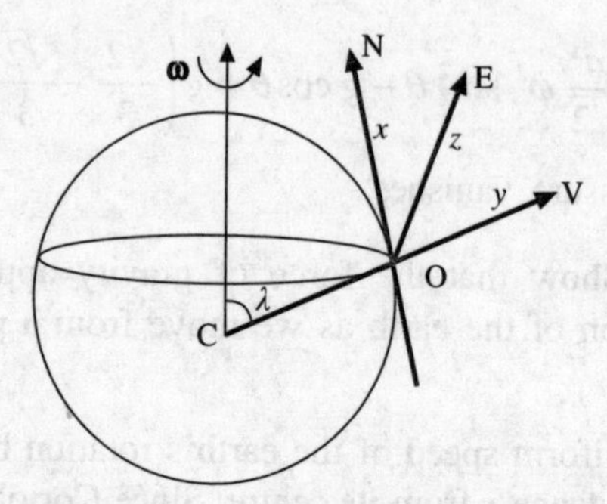

Figure 10.6 Co-ordinate system *x-y-z* attached at the point O on earth's surface.

With Figure 10.6 in view, we can express the equations of motion given in Eq. (10.26) adapted for the present problem as

$$\left.\begin{aligned} \ddot{x} &= -2\omega\dot{z}\sin\lambda && \text{(a)} \\ \ddot{y} &= -g + 2\omega\dot{z}\cos\lambda && \text{(b)} \\ \ddot{z} &= 2\omega(\dot{x}\sin\lambda - \dot{y}\cos\lambda) && \text{(c)} \end{aligned}\right\} \quad \text{(i)}$$

These equations are to be integrated with the following set of initial conditions

$$\text{At } t = 0,\ x = y = z = 0 \text{ and } \dot{x} = 0, \dot{y} = V\sin\lambda \text{ and } \dot{z} = V\cos\lambda \quad \text{(ii)}$$

Integrating once and applying the second set of conditions above, we obtain

$$\left.\begin{aligned} \dot{x} &= -2\omega z\sin\lambda && \text{(a)} \\ \dot{y} &= -gt + 2\omega z\cos\lambda + V\sin\alpha && \text{(b)} \\ \dot{z} &= 2\omega(x\sin\lambda - y\cos\lambda) + V\cos\alpha && \text{(c)} \end{aligned}\right\} \quad \text{(iii)}$$

Eliminating z in between Eqs. (iiia) and (iiib), we get

$$\dot{y} = -gt - \dot{x}\cot\lambda + V\sin\alpha. \quad \text{(iv)}$$

and on integration and applying the initial condition $y = 0$ at $t = 0$, above equation yields

$$y = -\frac{1}{2}gt^2 - x\cot\lambda + Vt\sin\alpha. \quad \text{(v)}$$

Next, substituting for $\dot{z}$ from Eq. (ia), and for y from Eq. (v) in Eq. (iiic), neglecting $O(\omega^2)$ terms and simplifying, we get the differential equation determining x as

$$\ddot{x} = -2\omega V \cos\alpha \sin\lambda.$$

Using the initial conditions $x = \dot{x} = 0$ at $t = 0$, we get

$$x = -\omega V (\cos\alpha \sin\lambda) t^2. \qquad \text{(vi)}$$

Substituting the above value of x in Eq. (iv), we get

$$y = -\left(\frac{1}{2}g - \omega V \cos\alpha \cos\lambda\right)t^2 + Vt \sin\alpha.$$

Now the equation $y = 0$ has two roots 0 and

$$T = \frac{2V \sin\alpha}{g - 2\omega V \cos\alpha \cos\lambda} \simeq \frac{2V \sin\alpha \left(1 + \dfrac{2\omega V}{g} \cos\alpha \cos\lambda\right)}{g}. \qquad \text{(vii)}$$

The missile will strike the horizontal plane through O (the plane $y = 0$) at the time T given by above expression. Hence, the perpendicular distance X of the strike point from the plane of projection is obtained on substituting $t = T$ from Eq. (vi) thus, we

$$X = -\frac{4\omega V^3 \sin^2\alpha \cos\alpha \sin\lambda}{g^2}.$$

Since x-axis points towards the north, the negative sign indicates the southward direction of the strike point.

PROBLEMS

1. A particle slides on a smooth helix of angle α under a force to a fixed point on the axis equal to $m\mu$ (distance). Show that the reaction of the curve cannot vanish unless the greatest velocity of the particle is $a\sqrt{\mu} \sec\alpha$.
2. A particle moves over an elliptic helix with constant angular velocity and with constant axial velocity. Show that its velocity and acceleration cannot be orthogonal except when the helix is circular.
3. A particle is attached to one end of a string of length l, the other end of which is tied to a fixed point O. When the string is inclined at an acute angle α to the downward-drawn vertical the particle is projected horizontally and perpendicular to the string with velocity V; find the resulting motion.
4. A heavy particle moves in a smooth sphere; show that, if the velocity is that due to the level of the centre, the reaction of the surface will vary as the depth below the centre.

5. Prove that if particles move on a right circular cone under no forces, the projection of their paths on a plane perpendicular to the axis are similar curves of the type $r \sin n\theta = c$, whatever be their initial velocities.
6. A particle moves on the inner surface of a smooth cone, of vertical angle 2α, being acted on by a force towards the vertex of the cone, and its direction of motion always cuts the generators at a constant angle β; find its motion and the law of force.
7. A particle moves on a smooth sphere under no force except the pressure of the surface; show that its path is given by the equation $\cot\theta = \cot\beta\cos\varphi$, where θ and φ are its angular co-ordinates.
8. If a particle moves on the inner surface of a right circular cone under the action of a force from the vertex, the law of repulsion being $m\mu\left[\dfrac{a\cos^2\alpha}{R^3} - \dfrac{1}{2R^2}\right]$, where 2α is the vertical angle of the cone, and if it be projected from an apse at a distance a with velocity $\sqrt{\dfrac{\mu}{a}}\sin\alpha$, show that the path will be a parabola.
9. Three masses m_1, m_2 and m_3 are fastened to a string which passes through a ring, and m_1 describes a horizontal circle as a conical pendulum while m_2 and m_3 hang vertically. If m_3 drops off, show that the instantaneous change of tension of the string is $\dfrac{g\,m_1\,m_3}{m_1 + m_2}$.
10. A particle moves on the surface of a rough circular cone under the action of no forces. It is projected with velocity V at right angles to a generator at a distance d from the vertex. Show that, when it has moved through a distance s, its velocity is given by $\log\left(\dfrac{V}{v}\right) = \dfrac{\mu s}{\left\{\tan\alpha(s^2 + d^2)^{\frac{1}{2}}\right\}}$, where μ is the coefficient of friction and α the half-angle of the cone.
11. A particle projected horizontally along the smooth surface of a sphere of radius a at the level of the centre. Prove that the motion is confined between two horizontal planes at a distance $\dfrac{(\sqrt{5}-1)a}{2}$.
12. A particle falls down a smooth tube in the form of helix. Find the pressure when it has fallen through a height h, if the axis of the helix is vertical and the particle is descending with uniform velocity v.
13. A particle moves on a smooth cone under the action of a force to the vertex varying inversely as the square of the distance. If the cone were developed into a plane, show that the path becomes a conic section.

14. A particle moves on the surface of a smooth sphere and is acted on by a force in the direction of the perpendicular from the particle on a diameter and varying inversely as the cube of distance. Show that it can be projected so that its path will cut the meridian at a constant angle.

15. A particle is projected northward with speed V at an elevation α from a point of the earth's surface in north latitude λ. Prove that the approximate deviation of the particle from the vertical plane of projection at time t is

$$\frac{1}{3}\omega t^2\,[gt\cos\lambda - V\sin(\alpha-\lambda)].$$

16. A projectile is fired towards the east at an angle of elevation α to the horizontal with speed V relative to the earth from a point in the northern hemisphere at latitude λ. If angular speed of rotation of the earth is ω. Neglecting terms of $O(\omega^2)$, show that the maximum height attained and the range are given by

$$\frac{V^2\sin^2\alpha}{2g}+\frac{\omega V^3\sin^2\alpha\cos\alpha\cos\lambda}{g^2},$$

and

$$\frac{V^2\sin 2\alpha}{2g}+\frac{4\omega V^3\sin\alpha(3-4\sin^2\alpha)\cos\lambda}{g^2}.$$

17. A bomb is dropped from a height h above a point O on the earth's surface whose latitude is λ. The earth is rotating with constant angular velocity ω. Taking O as origin and setting x and y axes (in the horizontal plane through O) along south and east directions, show that the bomb will hit the ground approximately at the point

$$\left(\frac{\omega^2 ha\sin\lambda\cos\lambda}{g},\ \frac{2\omega\sqrt{\left(\frac{2h^3}{g}\right)}\sin\lambda}{3}\right)$$

[*Hint:* Consider Eq. (10.24) taking into account both the *Coriolis force* and the *Centrifugal force*, and set the z-axis in the vertical direction]

18. A thin straight hollow tube is always inclined at an angle α to the upward drawn vertical, and revolves with uniform velocity ω about a vertical axis which intersects it. A heavy particle is projected from the stationary point of the tube with velocity $\dfrac{gt\cot\alpha}{\omega}$;. show that in time t it has described a distance $\dfrac{g\cos\alpha\,[1-\exp(-\omega\sin\alpha\cdot t)]}{\omega^2\sin^2\alpha}$.

[*Hint:* Here, we have $\theta = \alpha$ and $\dot{\varphi} = \omega$; these provide $\dot{\theta} = 0$ and $\varphi = \omega t$. Using these values the equation of motion in the R-direction may be expressed as

$$\ddot{R} - R\omega^2 \sin^2 \alpha = -g \cos \alpha. \tag{i}$$

Integrating Eq. (i) under the initial conditions $R = 0$, $\dot{R} = \left(\dfrac{g \cot \alpha}{\omega}\right)$ at $t = 0$, we get the required answer. (Draw the figure and work out)].

19. A smooth helix (radius a and angle α) is placed with its axis vertical and a small bead slides down it under gravity; show that it makes its first revolution from rest in time $2\sqrt{\dfrac{\pi \cdot a}{g \sin \alpha \cos \alpha}}$.

[*Hint:* $x = a \cos \varphi$, $y = a \sin \varphi$, $z = s \cos \alpha$

Thus, $ds = \sqrt{a^2\, d\varphi^2 + dz^2} = a \operatorname{cosec} \alpha\, d\varphi$, which on integration gives $s = a\varphi \operatorname{cosec} \alpha$. Using this, energy equation provides

$$dt = \frac{ds}{\sqrt{2gs \cos \alpha}}$$

$$= \frac{d\varphi}{\sqrt{2ag\varphi \sin \alpha \cos \alpha}}.$$

Integrating and imposing the condition where for a complete revolution φ changes from 0 to 2π, we get the required time.]

20. A particle moving over a paraboloid of revolution under a force parallel to the axis crosses the meridian at a constant angle α; show that the force varies inversely as the fourth power of distance from the axis.

[*Hint:* Since the external force $-F$ is parallel to the axis, the azimuthal equation of motion provides

$$\rho^2 \dot{\varphi} = h, \tag{i}$$

which is a constant. On the parabolololoid $\rho^2 = 4az$; this provides on differentiation with respect to time

$$\dot{z} = \left(\frac{\rho}{2a}\right)\dot{\rho}. \tag{ii}$$

Next, from geometry on the path curve, we have $\rho\left(\dfrac{d\varphi}{dz}\right) = \tan \alpha\, [(ds)^2 = (d\rho)^2 + (dz)^2]$ which together with Eq. (i) provides

$$\dot{z} = \frac{h \cot \alpha}{\sqrt{4a^2 + \rho^2}}. \tag{iii}$$

Hence, the force is

$$F = \ddot{z} + \left(\ddot{\rho} - \rho\dot{\varphi}^2\right)\frac{2a}{\rho} = -2a\,\frac{h^2}{\rho^4} \operatorname{cosec}^2 \alpha$$

Index